Steam Laundries

Johns Hopkins Studies in the History of Technology

MERRITT ROE SMITH,
Series Editor

Arwen P. Mohun

STEAM LAUNDRIES

Gender, Technology, and Work in the United States and Great Britain, 1880–1940

The Johns Hopkins University Press

Baltimore & London

Printed in the United States of America
on acid-free paper
1 3 5 7 9 8 6 4 2

The Johns Hopkins University Press
2715 North Charles Street
Baltimore, Maryland 21218-4363
www.press.jhu.edu

Library of Congress Cataloging-in-Publication Data
will be found at the end of this book.

A catalog record for this book is available
from the British Library.

ISBN 0-8018-6002-4

Contents

List of Illustrations

Facing page 1

The cover of *Scientific American*, 14 January 1893, depicting washers, centrifugal extractors, mangles, and soap-making techniques used in New York's Empire Steam Laundry.

Acknowledgments

Every book has its own history. An important part of the history of this book lies in the debts, intellectual and personal, I have acquired while writing it. Carroll Pursell, my dissertation advisor, first suggested the topic. He enthusiastically supported my efforts to turn a potentially narrow case study into something broader and provided an important example through his own wide-ranging interests and vibrant intellectual curiosity. Other committee members, Michael Grossberg, Jan Reiff, and Angela Woollacott, introduced me to their respective fields. I was slow to grasp what they tried to teach me but hope some understanding shows in the pages I wrote after I left their tutelage. Early on, a ten-week fellowship at the Smithsonian's National Museum of American History proved invaluable. In particular, Steve Lubar's guidance, friendship, and good humor helped convince me that laundries were worth studying and enabled me to get the project off the ground.

Had I not come to the University of Delaware to teach, I might never have realized that I was writing a book about the history of industrialization. Undergraduate and graduate students as well as colleagues will recognize some of the ideas I worked out in conversation with them. Conferences and seminars at the Hagley Museum and Library's Center for the History of Business, Technology, and Society at the Hagley Museum and Library provided an invaluable ongoing education in many of the subject areas this book encompasses.

Over a period of more than two years, Nina Lerman, Ruth Oldenziel, and I grappled with putting together a special issue of *Technology and Culture* on the use of gender analysis in the history of technology. Because of those conversations, this is a much better book. Nina once suggested that we were both part of a larger “thought collective.” I’m particularly grateful for her role in that collective, for her friendship, and for her ability to find the meaning in my cryptic rough drafts.

Ava Baron, Mike Grossberg, and Roger Horowitz all read chapters at critical junctures. Phil Scranton gave good advice on writing a prospectus when I needed help. Judy McGaw provided extremely valuable comments on the manuscript as a whole. Angie Hoseth and Ritchie Garrison helped with the illustrations.

This book would also not have been possible without the financial support I received from various institutions. In addition to the Smithsonian Institution, the Newberry Library and the American Association of University Women helped finance research and writing time. An Eva L. Pancoast Memorial Fellowship funded a summer’s research in London. A small grant from the University of Delaware allowed me to do archival research at the Wisconsin Historical Society and the Labor-Management Records Center at Cornell.

While equally important, some debts defy conventional categorization. Pieter Bergman, dear friend and constant enthusiast, died before the book was finished but left me a legacy that aided in its completion. Len Marsak convinced me by example that being a historian is meaningful work and gracefully helped me on my way when I decided not to become an intellectual historian.

As far back as anyone can remember, the women in my family have worked outside the home (and without doubt, inside their homes as well). This book is for them. I became a historian partly because my family encouraged me to find not just a job but a vocation. For that gift and many others, my heartfelt gratitude to Joyce Palmer Mohun, Martha Palmer, Rick Mohun and Susan Mohun, and my sister Rowena. Finally, Erik Rau arrived in my life after the laundresses and laundrymen, reformers, and frustrated customers were already firmly ensconced. He has been constant in believing in this project, contributing his skills as historian and editor when I’d exhausted my own, reminding me that I am indeed the lucky one.

Steam Laundries

SCIENTIFIC AMERICAN

A WEEKLY JOURNAL OF PRACTICAL INFORMATION, ART, SCIENCE, MECHANICS, CHEMISTRY, AND MANUFACTURE

Vol. LXVIII, No. 2.

NEW YORK, JANUARY 14, 1893.

INTRODUCTION

Close to Home

Laundry is a problem that refuses to go away. In an era in which many other traditionally domestic processes have been displaced from the home, the side-by-side washer and dryer are a persistent symbol of middle-class economic well-being. For the less affluent, spending time in the communal laundromat remains the norm. This, however, was not always the case. During the course of the nineteenth century, entrepreneurs and machinery manufacturers tried to remove laundry work from the domestic sphere and transform it into an industrial process. Beginning in the 1880s, increasing numbers of consumers chose "steam laundries," as they were most often called, over the washerwoman, the Chinese laundry, the washtub, and the sadiron. And growing numbers of women sought work in these laundries. This book is about that transformation in the United States and Britain, about the forms it took, and about its eventual failure as laundry work went back into the home after World War II.

The history of the laundry industry, particularly the process through which the relationship between home and work was reconfigured, offers an alternative perspective on the experience of industrialization. Laundries, even highly mechanized ones, were characterized by a persistent identification with domesticity and with women. Their customers were also their competitors. Their workforce was mostly female. Ubiquitous in urban areas, they remained small-scale rather than becoming centralized. Their peculiar qualities connect frequently separated analytical categories such as technology and culture, consumption and production, and domesticity and industrial processes. In a multitude of ways, laundries were close to home.

The overarching narrative of this book provides a detailed answer to the question of how and why laundry work became industrialized and then returned to the home. It argues that the rise and decline of the industry, as well as the shapes the industry took during the years in between, resulted from a dialectic between what was technologically possible (for both laundries and their competitors) and what was culturally desirable. This trajectory was a product of choices by consumers, owners, workers, and reformers. Close examination of this story reveals that cultural ideas, particularly about cleanliness, domesticity, and gender roles, played an extremely important part. It also became clear that commercial laundries suffered substantial technological limitations, which were acceptable to a critical mass of consumers only as long as viable alternatives did not exist.

Focusing on the producer and consumer of laundry services tells only part of the story. Although working-class women occasionally sent washing to laundries, the industry more often played a role in their lives as a source of employment. They approached the technology of commercial laundries not as a consumer choice but as a means of earning a living and as an environment (often hot and dangerous) in which they spent many of their waking hours. In turn, laundry owners depended upon the availability of large numbers of women with specific technological knowledge who could be hired at relatively low wages. The dynamics of the class and gender system were every bit as important as technological innovation in determining the industry's viability.

Workers and their advocates played a fundamental role in shaping the industry. Through unionization and state regulation, they helped reshape the terms of competition between laundry owners and ultimately the cost of services for consumers. Mechanization became a strategy to avoid the demands of skilled workers as well as a way to create excess laundering capacity after protective legislation had eliminated overtime and nightwork. The incursion of the state was not entirely unwelcome to laundry owners. Factory inspections and minimum wage rates drove the smallest and most marginal operators out of business, promised some measure of stability, and controlled competition.

Employers, consumers, workers, and their advocates (referred to throughout as "reformers") are the important actors in this story. All these people had a stake in how the technological process of getting clothes clean was organized. In a sense, they formed a set of overlapping communities, both local and international in character. As places where all these people interacted on a day-to-day basis, individual laundries were the industry's most concrete and intimate manifestation. Laundries also fit into neighborhoods and cities as sources of employment, as places to send dirty laundry and fetch clean clothes, as symbols of industrialization occupying streetcorners, and as annoying sources of smoke and dirty water. By the late nineteenth century, people had begun to talk about the "laundry industry." This referent, which lacked any precise spatial definition, comprised trade associations and union meetings, the discourse of reformers, and the laws that regulated laundry owners' businesses.

Many tools and approaches are useful in opening up various aspects of the industry to analysis. Without understanding the dynamics of gender roles, relationships, and meanings, the economic and cultural dynamics of this story make little sense. The importance of gender difference took on a multitude of forms, including the sexual division of labor, the relationships between male laundry owners and female consumers, and the way in which gendered meanings were attached to the process of laundering itself. Gendered identities and relationships were also complicated by race, ethnicity, and class. For instance, laundry work, consumption, and, in Britain, management all carried specific class as well as gender connotations.

In the laundry industry, the cultural and technological, the domestic and industrial, intertwined in complex and sometimes contradictory ways. Because most laundries did not produce material objects, laundries have long been described as a service industry. This categorization helps explain the relationship between laundries and their customers, but it misrepresents the essential technological and organizational logic of the steam laundry. As far as laundry owners and their workers were concerned, steam laundries were factories. Laundry owners on both sides of the Atlantic strove to emulate Henry Ford, not the banks, insurance companies, or department stores often included under the ambiguous umbrella of "service industry." Contemporary observers who drew parallels between the cottage spinner replaced by the cotton mill and the washerwoman displaced by the steam laundry understood what the industry was trying to do. Centralization, economies of scale, mechanization and deskilling, division of labor, organization, and control of information were all critical strategies for mechanizing laundry work.

This book might best be considered an Anglo-American history with a strong comparative component.[1] At one level, laundries remained profoundly local businesses. Laundry owners competed mostly within individual towns and cities. Local prejudices and local economies governed patterns of consumption. Local labor markets also defined who would work in a laundry and whether laundries could compete with washerwomen and Asian newcomers. At another level, however, the industry was a transatlantic phenomenon. Continual exchange of machinery and techniques informed its history, as did two transatlantic conversations: one between British and American laundry owners and another between British and American reformers. While the exchanges were mutual, machinery and technical knowledge tended to flow from the United States to Great Britain while ideas and practices about reform traveled in the opposite direction.

All this exchange resulted in some manifest similarities in technology, division of labor, and business organization. However, differences in context–infrastructure, ethnic and racial composition, legal systems–resulted in notable dissimilarities. Most conspicuously, British reformers worked within a constitutional framework that allowed effective regulation of industry much earlier than was possible in the United States. The British industry was relatively homogeneous in its ethnic makeup; American laundries were often ethnically and racially diverse. In Britain, class distinctions took on a comparatively more important role. In the United States, regional differences also had enormous effect on the shaping of the industry and its regulation.

In their own time, British and American laundry owners and writers for trade journals also liked to emphasize these dissimilarities: Americans, they wrote, were quicker to mechanize and shoddier in their construction of machinery. They tended to build large and discard quickly. However, other evidence suggests a remarkable degree of consistency in what might be termed *best practice* on either side of the Atlantic. Highly capitalized laundries in London and Chicago resembled each other far more than they did their small, local competitors. This was a result not just of parallel development but also of sustained buying, borrowing, and visiting across the Atlantic. As economic entities, laundries are not like railroads. Householders generating their own electricity or running a handcar service in competition with the Pennsylvania Railroad might seem ludicrous, but the analogous situation was not laughable at all to laundry owners.

Laundry owners' nationalistic rhetoric served a number of complex cultural functions. Such observations were sometimes offered solely for purposes of amusement, but they more often served to reinforce a sense of national identity. One of the paradoxical effects of modernization and industrialization has been to erase differences in material life while constructing ideologies that reinforce ideas of national uniqueness.

I have used comparison where it seems most useful: to highlight factors that might go unnoticed in a single national context; to illustrate the existence of continuous transatlantic exchange; to show the importance of structural factors such as race; and to get at some of the elusive aspects of cultural difference. When it seems necessary, I have also diverged from the nation-state as the basis of comparison. The later chapters, in particular, recognize the importance of regionality, especially North-South distinctions in the United States.

It should also be said that within the tidy term *laundry*, bandied about in census records and trade journals, lies an array of very different kinds of business. Laundries could (and still can) be found attached to a variety of institutions: hospitals, hotels, lunatic asylums, orphanages, as well as manufacturers of shirts, towels, and other items made of cloth. The dynamics of such specific laundries lie outside the scope of many of my sources and, therefore, this story. On the other hand, laundries that specialized in the "family wash" or the "bachelor bundle" are generally impossible to distinguish from linen and uniform services: many laundries did both. I have left that aggregation intact, in part because of the nature of my source material and also because it is telling. Hotels and restaurants tried to recreate the comforts of home on a grand scale. They too are inventions of this industrializing age and its accompanying logic. This book touches only peripherally on the history of so-called Chinese laundries. For most actors in this story, Asian laundry owners and their employers were the "other"–unemployable, unregulatable, and unacceptable as trade association or trade union members. Consumers regularly crossed the line, however, ignoring racial protocols for services that were cheaper and sometimes less destructive to clothing than those offered by steam laundries.

⌘

In the past twenty or more years, feminism and its scholarly manifestations, women's history and gender analysis, have transformed the ways some scholars approach the history and sociology of technology. This book is, in part, a product of the arguments they have made, namely that gender is important in shaping technology and technology in shaping gender; that the study of baby bottles and sewing is as important as the study of engineering and large machines made out of metal; and that women, whether housewives or laundresses, are technological actors, not just passive victims of technology.[2]

Approaching laundries from the outside, this book is about businessmen and consumers; from the inside, it is about women and work. A rich literature, generally identified with labor and the social history of women and aimed at making visible the gendered connections between home and work, has influenced my overall analysis. More than a generation of scholarship has shown how inseparable the patterns of women's work lives are from their roles as mothers, housewives, and caretakers. Other historians have demonstrated the ways in which gender ideology has shaped notions of skill and sexual division of labor and has linked the devaluation of women's work in the labor market with its identification with domesticity.[3] The rhetoric of reform, relations between consumers, workers, and managers, and even the spacial construction of workplaces all have been shaped in part by gender.[4]

While no adequate history of the laundry industry exists, I am not the first scholar to have given thought to its peculiar connection to the home.[5] In *More Work for Mother,* Ruth Schwartz Cowan describes steam laundries as a technological "road not taken," using them as one of many examples to illustrate why some alternatives to housework succeeded while others failed. Cowan's explanation for the laundry phenomenon focuses primarily on the values attached to the home and homemaking. She argues that the consumers' choices to reject communal and commercial alternatives might best be explained by understanding that "most people will still opt for privacy and autonomy over technical efficiency and community interest."[6] In a sense, this book mirrors Cowan's work, looking at one aspect of the home in the industrial revolution rather than the industrial revolution in the home.

As previously stated, this book has been conceptualized partly as a study of social communities organized around a particular technological process. This approach has the virtue of joining social relationships and industrial processes, two subjects that other forms of analysis isolate. Community studies have been important to the history of technology because they put technology in context.[7] I have, however, expanded the notion of community beyond the spacial constraints of the nineteenth-century industrial village. Even in the prosaic world of laundries, new technologies and new social forms stretched the traditional boundaries of time and space.[8]

Finally, while many different kinds of historians write about aspects of industrialization, the task of summing up has most often been left to economic historians. A few of them have made social, cultural, and technological history part of their approach, questioning and complicating a narrative that focuses only on the rise of large-scale, capital-intensive enterprise.[9] I do not attempt a grand synthesis, but rather intend to suggest some ways of writing industrial history that may eventually lead to a retelling of that larger story.

⁜

The first chapter sets up the argument that both functional technology and a suitable social and cultural context were necessary to make steam laundries economically viable. The industry originated in the nineteenth century owing to the availability of technology and the development of techniques to do laundry in a factorylike setting. At the same time, sooty skies and fouled water made it increasingly difficult to use traditional laundry methods in urban environments, and rising standards of cleanliness among the urban middle class helped create a market for laundry services.

Although commercial laundries existed as early as the 1850s, they were few in number and left little evidence in the historical record. Chapters 2–5 cover the period from 1880 to 1914, during which the industry began to grow rapidly and underwent a period of extensive mechanization. This period is also significant because it marked the emergence of a publicly recognized entity: the "laundry industry." In these same decades, the women constituting the majority of the workforce began to attract substantial public interest from reformers and social investigators.

Chapter 2 describes the industry as it existed between 1880 and about 1910 and introduces laundry owners as important protagonists in the story. I begin with a description of the beginnings of the American Steam Laundry in Newcastle-upon-Tyne in 1884. This story introduces the themes of periodization, the gendering of laundry ownership as male, the international trade in technology, and the problems of running a laundry. It also emphasizes the roles gender and race played in differences between the British and American laundry owners' communities.

In Chapter 3, I describe the technological processes and machinery involved in a typical laundry of the period, following the operation from beginning to end. This chapter illustrates why contemporaries often described laundries as factories in terms of both their technology and their organization of work. Goods entering laundries not only had to be washed and dried but also starched and ironed. To carry out these tasks, laundry owners gradually replaced the hand iron and tub with increasingly minute divisions of labor as well as a wide variety of specialized machinery. As laundries grew in scale, new systems of sorting, marking, and recordkeeping developed. My description also suggests that while laundry owners viewed the process in terms of production, workers (and their advocates) saw it primarily as an environment in which they spent much of their time. The chapter is also intended to acquaint the reader with the physical setting of a steam laundry and to suggest the kinds of technological resources and knowledge needed to do laundry work.

The women who worked in laundries are introduced in Chapter 4. I argue that the gender and class systems that made large numbers of women workers available at relatively low wages was as crucial to the existence of the industry as were the technological innovations. What it meant to be a woman and a laundry worker changed over time and with increasing organization and mechanization. I compare the different cultural meanings the British and Americans gave to traditional laundresses and suggest some reasons for the slower pace of mechanization in Britain. I also show the complex ways in which the gendering of domestic laundry work was translated into an industrial workplace.

Chapter 5 follows the intertwined efforts both to unionize laundry workers and to use legal means to ameliorate working conditions. In this period, laundry workers were frequently used as examples of wage labor's socially detrimental effects on women. Laundry workers were selected for this role because of the familiarity of laundries to many middle-class consumers and because of the symbolic meanings of femininity, domesticity, and cleanliness attached to laundry work. While some laundry owners fought against efforts by workers and reformers to reshape conditions, others recognized that organization and regulation could be used to control the behavior of both workers and competitors. Finally, I use comparison to explore the differing effects of culture, the role of the state, and traditions of labor organization in shaping this process on either side of the Atlantic.

The chapters in Part 2 deal with the industry as it matured and later began to decline, during the interwar period. In Chapter 6, I argue that, by the 1920s, the laundry industry had reached a state of technological maturity. However, laundry owners found it increasingly difficult to find ways to make their methods cheaper, more efficient, and more reliable. Their problems were aggravated by increased regulation, unionization, and the availability of new electric washing machines designed for the home. They hoped in vain for a Henry Ford of the laundry industry: instead they settled for new types of machinery-intensive services. They also used marketing and advertising efforts to bridge the gap of understanding between consumers and laundry owners. British laundry owners, meanwhile, looked to the United States for solutions, straining to adopt both the ideology and the methods of American business to a British context.

In the United States, the interwar period was also marked by the influx of large numbers of African American workers. On the shopfloor, segmentation created a two-tiered workforce, divided primarily along racial lines. Wages, division of labor, and workplace culture were all affected. Race became a central factor in efforts to create effective unions and to reform the industry through legislative means. Chapter 7 provides a short interlude, which introduces these African American workers and describes their entry into the laundry workforce.

In this period, state regulation became an increasingly important factor in the industry. In Chapter 8, I compare two interwar efforts to regulate the laundry industry: the National Industrial Recovery Act (NIRA) in the United States and the Trade Board Acts in Great Britain. I show how ideas about industrial reform crossed back and forth across the Atlantic, shaping both programs; how efforts designed for large-scale industries such as textiles and steel became problematic when applied to the laundry industry; and how race became a key issue in the efforts to create a laundry code in the United States. The chapter picks up threads from Chapter 5, showing the symbolic role that laundry workers continued to play in debates about women in industrial work. In keeping with chapters from the previous section, it also shows how some laundry owners welcomed regulation as a means of stabilizing the industry and eliminating some forms of competition.

Most laundry owners experienced regulation not through the policymaking process but through later implementation and enforcement. Chapter 9 compares efforts to implement the NIRA and Trade Board Acts, illustrating how those efforts were cast to fit British and American cultural differences. Laundry owners in turn made choices about whether or not to cooperate based on their understanding of the power of the state and their relationship to it.

In Chapter 10 I compare two American laundry unions of the 1920s and 1930s. The membership of Local 300, a New York union, was primarily composed of African American women. Milwaukee Local 360 was a union of white male drivers–the highest paid, most independent employees of the laundries. I examine how gender and race affected the organization of these unions as well as workers' perceptions of themselves and their relationship to their unions. By describing the post–Wagner Act accord reached between laundry owners and managers, Chapter 10 finishes the comparative story begun in Chapter 8, showing that these Congress of Industrial Organization (CIO) unions created private versions of the Trade Boards through various organizations and mechanisms, including bargaining procedures and joint boards.

In the end, what ultimately led to the decline of the industry was the threat of home washing machines, and so with Chapter 11 I return to the theme of technological choice introduced in the first two chapters. In increasing numbers, consumers began to purchase washing machines rather than take their clothes to the laundry. I describe the struggle between washing machine manufacturers and laundry owners for the business of consumers and suggest some of the reasons why laundries eventually lost that battle, as laundry work returned to the home.

PART I

Origins and Growth of an Industry, 1880–1920

1

Technical and Cultural Origins

WHERE AND WHEN DOES THE HISTORY OF AN INDUSTRY BEGIN? How and why does a domestic task become an industrial process? Like other human endeavors, the laundry industry has its own myths of origin. A version favored by Americans describes an entrepreneurial miner in the California gold rush who hooked up a jury-rigged washing machine to a donkey engine, committing an act of invention in that most American of settings, the frontier.[1]

Why did it begin? In this story, because the miners lacked women to wash their clothes. Thus the first laundryman created a technological solution to a social problem. In both American and Western European cultures, laundry work was traditionally one of the most powerfully gendered of domestic tasks. Whether in the twentieth-century industrial laundry of industry historian Fred De Armond's experience or the kettle and streamside of the frontier miner's world, women did the washing. Without women, the wash most often did not get done.[2]

The story of the frontiersman is probably apocryphal: prototypes of machines commonly used in early-twentieth-century laundries substantially predate the California gold rush. At the end of the eighteenth century, at least one English textile manufacturer created a system of steam-powered machines to wash and dry laundry in institutional settings. Eager entrepreneurs, however, did not immediately seize on these techniques. Commercial viability required an appropriate cultural and technological context. The industry first prospered not on a distant, masculinized frontier but rather in urban centers. Rising standards of cleanliness, concentrations of potential customers, and the creation of urban water systems aided its growth, while the degraded environment of cities made traditional clothes-washing methods less workable. Women played an essential role as customers, workers, and competitors. Culturally and physically, the industry grew up close to home, not in the distant and undeveloped American West.

If the facts are inaccurate, the story of the clever miner has a kind of seductive cultural logic. It begins with a moment of invention. This is the familiar tale of Yankee ingenuity in which an amateur inventor takes a piece of familiar domestic technology and alters it subtly but with enormous ramifications.

The British version of the industry's origin lacks frontier trappings, substituting instead the symbolism of heraldry and the authority of the patent office. Like its American counterpart, however, the story encompasses domestic and gendered linkages. In an official industry history, Ancliffe Prince presents a coat of arms created for the Company of Launderers of the City of London (founded in 1960, despite the ancient-sounding title) as a heraldic agglomeration of the industry's origins. The left-hand supporter of the coat of arms, a female figure, carries a dolly: a long wooden pole with cross pieces at the end, traditionally used by British laundresses to agitate clothes in a wash tub or boiling copper. The coat of arms also includes the image of a 1782 washing machine taken from patent drawings (the Sidgier machine) "between three antique flat irons fired proper."[3] Prince shows no apparent awareness of the irony of this largely male fraternity adopting as its symbol the traditional tools of the same female workers displaced by mechanization.

De Armond's device of altering a domestic washing machine for commercial use and Prince's appropriation of the laundress's traditional tools both shape the narrative of technological change into a story of succession and progress. Invention replaces the domestic with the industrial, the feminine with the masculine. These narrative strategies reify the laundrymen's authority as men in charge of a process that had been identified with women, and such tales fit neatly within the conventions of technological storytelling.[4]

Closer examination reveals that industrialization spurred the development of both domestic and commercial laundry technologies, along parallel but not always intersecting paths. For instance, inventors and manufacturers introduced hundreds of home washing machines during the nineteenth century, but the design of these new devices diverged significantly from their commercial counterparts. In *The Growth of Industrial Art*, a volume commemorating the 1889 centennial of the patent office, twelve examples of traditional techniques and patent machines illustrate some of the differences. Most domestic devices strove to imitate the motion of the human hand scrubbing clothes.[5] Number five, the sole exception in the illustration, is a wash boiler. All these machines are shown being operated by women, reinforcing their domestic connection. Number seven, the only commercial washer, is depicted without an operator because it is driven by a belt attached to a steam engine–an arrangement unobtainable in most homes. Like other contemporary commercial machines, and those used today, it consists of a rotating drum that moved water through clothes by tumbling. Makers of such devices adapted designs from commercial textile processing rather than scaling up home washers.[6]

Focusing on a single device, the washing machine, gives a misleading impression that doing laundry is synonymous with washing. Whether in a kitchen or as a commercial enterprise, laundering was a multistep process involving a variety of tools and machines. Most items also had to be wrung out, dried, starched, and ironed. As with domestic and commercial washing machines, other kinds of technology often took differing forms. At the end of the nineteenth century, housewives and washerwomen could purchase mass-produced coppers or boiling tubs, washboards, wringers, mangles, and clotheslines from catalogs and corner stores. Industrialization made these items cheaper, and enthusiasm for invention gave rise to a plethora of small improvements and patent gadgets.[7]

In contrast, the development of commercial laundry machinery followed the logic of factory production, dividing tasks into incremental components and incorporating elements of craft knowledge into the machine design. Other devices aimed to take advantage of economies of scale. For instance, hand ironing of flatwork was gradually replaced by huge cylinder irons (mangles) that could iron hundreds of sheets an hour.[8] Because the application of mechanical sources of power to do-

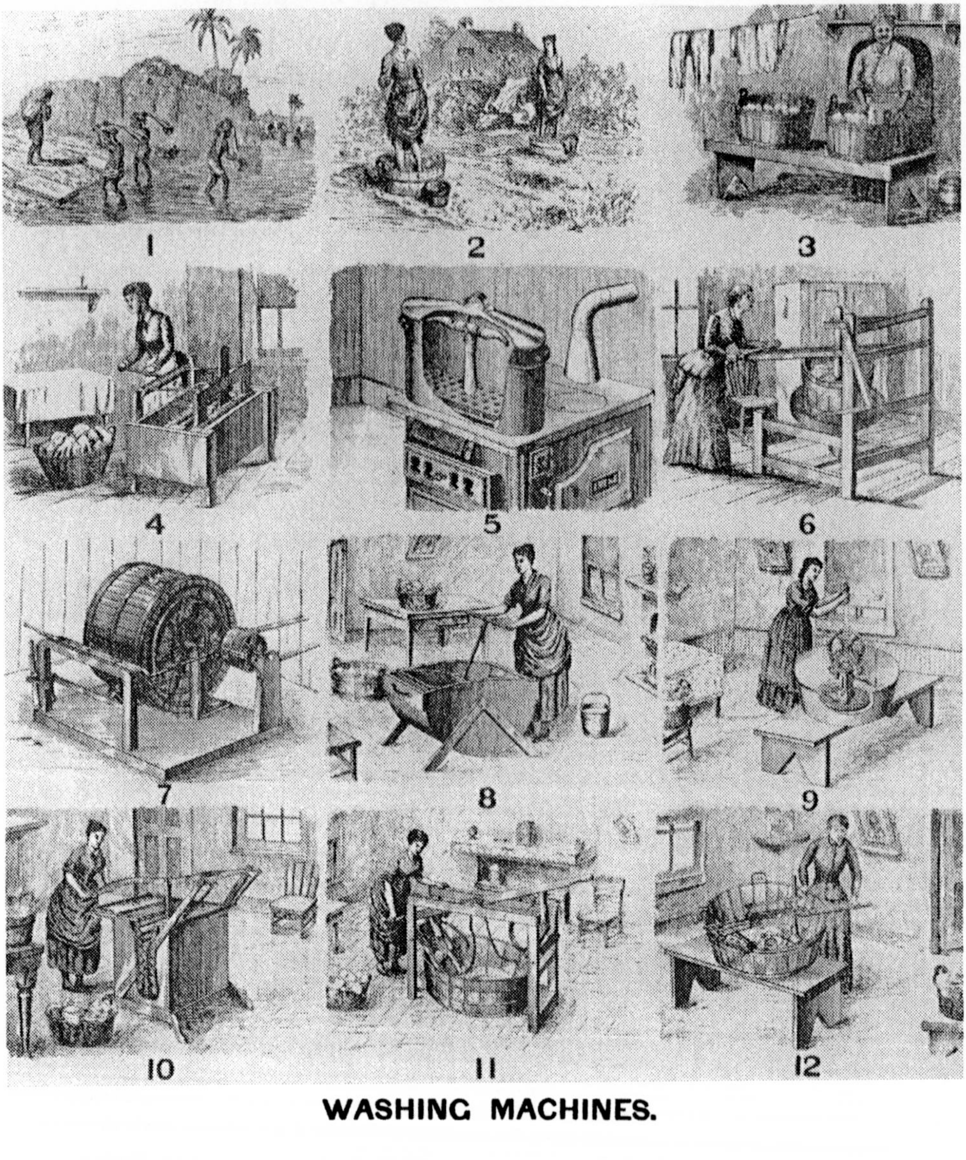

WASHING MACHINES.

Fig. 1. Most washing machines designed for home use employed rubbing rather than tumbling to move water through clothes. The gendered character of domestic laundry work is also emphasized in this 1892 illustration from Benjamin Butterworth's *The Growth of Industrial Art* (Washington, D.C., 1892), 84. Courtesy of Hagley Museum and Library.

mestic processes remained largely impracticable in ordinary households until the beginning of the twentieth century, industrial machines could not simply be scaled down for home use. Further, the division of labor that characterized large laundries and its accompanying specialization of machines to incremental tasks would not have made sense in the home or in one-woman commercial operations.

The tools available for domestic laundry work barely eased the back-breaking nature of the labor. Many housewives loathed this task above all others and actively sought alternatives to doing the wash at home.[9] Housekeeping advice books of the time give some sense of the onerousness of this process compared with the late-twentieth-century experience of automatic washers and dryers, electric irons, colorfast and permanent press fabrics, and specially formulated detergents and bleaches. Eliza Leslie's 1840 *Manual of Domestic Economy,* for instance, dedicated twenty pages to washing and ironing and an additional ten pages to methods of removing stains and grease from various fabrics.[10] In many nineteenth century households, "getting up" even a single family's wash took two days of exhausting labor.

Laundry day began by obtaining and heating water. In households without plumbing, water had to be hauled from the well, the creek, or the standpipe into the yard or the house. Leslie warned that "unless you have an abundance of water, it is impossible to get clothes clean."[11] Other sources also give some sense of the prodigious amount that had to be pumped, carried, and dumped out. Testifying before the Health of Towns Commission in 1843, one witness told members of Parliament that many urban laundresses hauled thirty pails of water (40–60 gallons), boiling half of that to accomplish their weekly washing.[12]

Leslie advised beginning the actual washing on Tuesday, rather than Monday as tradition had dictated. On Monday night, laundry could be sorted, stains treated, and particularly filthy items put to soak. Some garments even had their seams ripped out prior to washing, to be reassembled later. The following morning, each load was put into a tub of suds. Leslie recommended scrubbing each item on a washboard.[13] The dolly, or dolly peg (described by Prince), was another labor-saving device, used to punch and pound a tub or barrel full of laundry. During the nineteenth century, Americans and the British shared a belief that white goods were not really clean until they had been boiled. Therefore, cotton and linen goods were put into a copper or large caldron and heated over the fire after they had been washed.

If the laundress's back was not already aching after this washing stage, lifting sodden goods in and out of a wash tub for a series of rinses would guarantee it. Large items such as sheets, tablecloths, and blankets were particularly cumbersome, which was one reason these items were frequently sent out. Women lauded the appearance of patent wringers consisting of two parallel rubber rolls turned by a crank. Without such a device, each item had to be twisted by hand to remove excess water. The final step on the first day of washing was hanging wet laundry out to dry.[14]

In an age before permanent press, nearly all linens (unlike woolens) required ironing. The laundress prepared shirts, table linen, and other items by rubbing them with boiled starch. Flatirons, often weighing eight to twelve pounds each, were heated on a stove. Leslie and other advice writers recommended three or more irons for each woman ironing, so that time would not be wasted waiting for an iron to heat or an overheated iron to cool.[15] Ironing itself required both careful judgment to avoid scorching fabrics and physical strength to push and bear down on heavy irons.

Most domestic manuals did not linger over the fact that weather, inadequate tools, and cramped domestic spaces complicated this regime. Many women already knew that hot summer days made standing over a boiling copper or pushing a twelve-pound iron in the same kitchen as a lit woodstove a foretaste of hell. Rain and snow cooled overheated kitchens but meant that wet laundry had to be hung inside the house, where it soaked up household odors while contributing its own fetid smell to the indoor air. Constant immersion in hot water and harsh chemicals also took its toll on skin. At midcentury, many women still made their own soap out of animal fat and lye–a formula that was more than a little caustic.[16] Red, raw hands revealed a woman who did her own washing. Moreover, after the conscientious housewife spent two days on laundry work, there were only five more before the process began again.

Hiring a washerwoman was the traditional solution to these problems. However, using a hired washerwoman (whether she worked at her employer's home or took laundry away and returned with it) also posed problems for the housewife. Washing done in the home continued to disrupt domestic life, and middle-class women became increasingly wary of laundry done in worker's homes because of potential exposure to diseases. Either option was further complicated by all the tensions of finding, supervising, and keeping workers.[17] For the washerwoman herself, meanwhile, laundering offered a source of income but also meant heavy lifting and harsh chemicals all week long.

Because of the shortcomings of these domestic alternatives, housewives recognized the advantages of industrializing laundry work. Although the "bachelor bundle" created income for laundries catering to the domestic trade, single men bereft of women to do their washing were not commercial laundries' principal constituency—women were. Likewise, women, not men and their machines, continued to do most of the laundry work, even inside mechanized laundries.

Technological Origins

If the origins of the industry are synonymous with the origins of its technology, the American frontier, distant from the technology of the age and the means of reproducing it, makes little sense as a starting point. The British strategy of searching for patent records also provides limited help, since the public lives of many patent devices began and ended at the patent office. Other sources lead to technological precedents with a working existence, however. In 1819, Charles Sylvester ("engineer"), published one of those numerous nineteenth-century manuals of domestic economy, its short title: *The Philosophy of Domestic Economy; as Exemplified in the Mode of Warming, Ventilation, Washing, Drying, and Cooking.* In it he describes a laundry invented in 1796 by one William Strutt, Esq., and installed in the Derbyshire General Infirmary in 1810. Though flattery of an important man probably played some role in the writing of the volume (Strutt was a member of a prominent manufacturing family), Sylvester declared a humanitarian motivation for sharing information about Strutt's accomplishments. In particular, machinery would save women from such labor: "washing should be performed by a *machine*, much more fitted for extricating the dirt than the hands of the washer woman," he told his readers.[18]

The variety of technologies employed in the infirmary laundry bear a striking resemblance to the machinery commonly used in most late-nineteenth-century steam laundries. Sylvester's descriptions also tie them to their origins in eighteenth-century textile production. Instead of the washerwomen's traditional tools, the dolly and a tub, Strutt provided a washing machine consisting of a perforated cylinder divided into four compartments and enclosed in a watertight box. This washing machine, Sylvester informs the reader, was "adapted from the common washing wheel used by bleachers." It was applied with "great success in getting up and finishing stockings, previous to its use for linen" (stockings being one of the products of the Strutt-Arkwright mills). The operator should fill the box with water to a depth of four to five inches, letting in steam to heat the water. Once full, "turn [the cylinder] at speed to let clothes fall from one side to the other" instructs Sylvester. Although the accompanying drawing shows a hand crank on the side of the machine, Sylvester advises the reader that a six-horsepower steam engine installed in the basement turns the machine by means of a horizontal shaft.[19] Figures 2 and 3 contrast Strutt's machine with a Troy Laundry Machinery Company washer assembly room of 1911.

Similarities in drying and finishing devices also connect eighteenth-century textile production with later commercial laundries. Sylvester describes in great detail a drying closet: The closet has narrow compartments into which were slid "horses"–racks used to hang wet linen. Heat from a central furnace forced air through the closet. Sylvester notes that "these stoves have long been applied upon an extensive scale, to the drying of cotton yarn and calico, in the works of Messrs. Strutt; and they would not be less important to the paper maker, potter, sugar baker, &c." Other provisions of the wash house show an awareness of the shortcomings of domestic laundry methods and the burdens they imposed on the laundress. Sylvester indicates that "every wash house should be furnished with hard and soft water, by pipes, and with other pipes to convey steam, so as to render fire places unnecessary."[20]

Strutt was far from the solitary yankee inventor tinkering away with simple tools in a homely workshop. He had both the material resources and the technical knowledge to carry out this project. One of three sons of Jedidiah Strutt, a pioneering textile manufacturer and Richard Arkwright's partner, he was literally born into the epicenter of the Industrial Revolution. When William and his two brothers inherited the Strutt estate in 1797, the family holdings included four cotton mills in Belper, employing one-third of the inhabitants of the town, as well as a bleach works and an iron foundry at nearby Milford.[21] While his brothers tended the financial side of the business, William looked after the technical aspects of a rapidly evolving industry. Hence, he would have had frequent interaction with the mechanics of the family's Milford foundry, who could have turned out the necessary machinery or provided practical advice based on their experience with similar devices.

Strutt's interest in technology far transcended industrial innovation for the sake of profits. He was an avid inventor with a particular interest in improving conditions of daily life for the laboring classes.[22] As was true for many reform-minded ideas of Strutt's time, this aspiration took the form not of higher wages or improvements in the homes of workers but rather of institutional provisions: schools and hospitals, where paternalistic control could be maintained. This pioneering generation of entrepreneur-reformers saw technological innovation and scientific analysis as tools to ameliorate the ills of society (Strutt's friend Jeremy Bentham most notoriously). They were the children of both the Industrial Revolution and the Enlightenment–recipients of its ideology as well as its technology.[23]

Strutt considered the infirmary to be his masterpiece, taking pride in displaying in it "a whole range of mechanical genius." He shared not only his general enthusiasms but also his specific interests in washing machines and the like with the great men of the English Enlightenment. In 1811 he wrote Thomas Edgeworth concerning improvements in the washing machine. Correspondence with Bentham in the 1820s included a discussion of the design of boilers and heating systems for the infirmary. An 1803 patent specification states that "the second washing machine is an improvement on Bentham's machine," suggesting one of the Benthams may also have taken a turn at such "improvements."[24]

Strutt never patented any of his laundry machinery. Most likely, he could not make any claims to originality since there were already a number of eighteenth-century patents for dash wheel machines, most intended for use in the textile industry.[25] Moreover, if his goal was charity rather than profit, patents would not have facilitated the installation of laundries in hospitals and orphanages. It is difficult to establish a direct connection, either through patents or machinery manufacturers, between Strutt's machinery and the laundry machines that

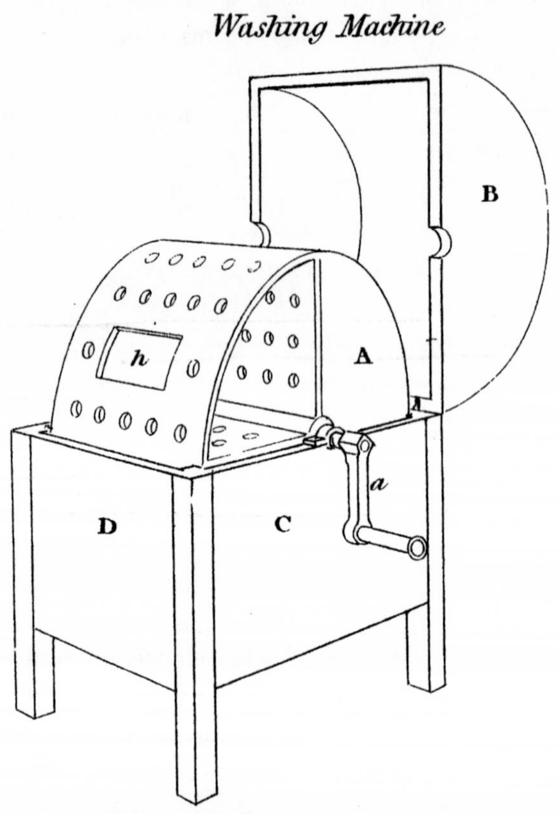

Fig. 2. William Strutt's 1810 washing machine shares the same basic design principles as the commercial washers commonly used a century later. Plates from Charles Sylvester, *Philosophy of Domestic Economy* (Nottingham, 1819). Courtesy of the Library Company of Philadelphia.

Fig. 3. The Troy Laundry Company illustrated its 1911 *Troy Laundry Guide* with this photo of the department in which washers, similar in design to Strutt's machine, were assembled at its Illinois factory. Troy Laundry Machinery Company, *Troy Laundry Guide* (Chicago, 1911), 58. Courtesy of Hagley Museum and Library.

began to appear at midcentury, but they bear a striking resemblance to each other. Sylvester hints, however, that the technology was adopted on a wider scale. He cites one other specific installation, the Pauper Lunatic Asylum in Wakefield, where "the laundry is similar to that at the Derbyshire infirmary, with the addition of machinery for turning the mangle, by which much labour is saved."[26]

Despite Strutt's prominence, no evidence survives to suggest that eager entrepreneurs immediately snatched up those ideas and set about making their modest fortunes. Americans, although eager to have the other fruits of British inventive genius through fair means or foul, also seemed to have ignored Strutt's institutional contribution to humanity, concentrating instead on laundry technology for domestic use.[27]

Given Britain's early industrialization and urbanization, one would expect that not only the first machines to mechanize laundry work but also their first commercial applications would appear on that side of the Atlantic. Indeed, at least one of Strutt's contemporaries, Sir John Dalrymple, published a pamphlet in 1802 detailing his scheme for using machinery to save labor costs in commercial laundry work. Dalrymple was less technologically astute than Strutt–his cumbersome technique involved piping steam through vats of dirty clothes or linen in bleaching solution over a period of hours–but he understood, as Strutt (or perhaps Sylvester) did not, the possibilities of mechanization for saving money and time. Dalrymple also recognized that urbanization provided the context to make mechanized laundry work economically viable. His device, he proclaimed, would be of particular advantage to London and other "great cities," where "washing costs triple of what it does in any other part of the kingdom." As later laundry owners' claims would echo, he also pointed out the sanitary advantages of machinery to its urban alternative, "where the washing is performed in garrets, often by careless unknown people, without drying places; and where there are no washing greens to spread clothes upon."[28]

Besides the cost and sanitary benefits, the utilization of machinery would also save work. "In washing by steam, in which the power of art does what the hands of washerwomen formerly did, 8,000 shirts can be washed in a day, which, in washing by hand, would have required 222 women, at 36 shirts to each woman," Dalrymple calculated. "There must therefore be a great gain on the acceleration of work, and consequently in the number of vessels, the extent of house-room requisite, the readiness to supply washing customers, and other consequences which arise from the quickness of work such as turning of money quick."[29]

Despite Dalrymple's efforts, during the first decades of the new century washerwomen still held a virtual monopoly on laundry services in London. However, at least one source suggests the beginnings of larger business. In 1834, the *Pioneer*, a very early labor journal, told its readers that washerwomen in the London neighborhood of Horseferry Road had declared a general strike. Their employers were described as "some of the great laundresses," suggesting that some entrepreneurial London laundresses had expanded their business into what might be characterized as factories without machinery.[30] Who those proprietors were and what their places of business looked like remains a mystery.

Henry Mayhew counted 51,289 people in London by 1850 earning their living by doing other people's laundry. He distinguished between three types of laborers: "laundry-keepers," "washers," and "manglers." "Laundry-keepers" constituted only 16,220 of the total, both their name and number suggesting a group of more substantial entrepreneurs.[31] Advertisements and listings in city directories, the appearance of subsidiary industries such as machinery manufacture, complaints from customers, and concerns of reformers all suggest that by midcentury laundries had begun to be publicly recognized as distinct commercial enterprises on both sides of the Atlantic.

It is more difficult to ascertain whether machinery was used and if so, what kind, how work was organized, and how many people labored in these places. Both the knowledge and technology needed to put a steam laundry together were widely available in the 1830s–one survey counted nineteen dying, bleaching, and calendering establishments using steam power in Philadelphia alone–but whether those technologies made the jump to washing laundry remains a mystery.[32] It is probable that, like the majority of small-scale manufactories of midcentury Britain and America, these businesses were workshops with only a few employees, without the capital–or the need–to mechanize.[33]

By the late 1850s, the term *steam laundry* had come into common usage. The 1859 *Ladies Philadelphia Shopping Guide and Housekeeper's Companion*, for instance, gave special attention to the "Philadelphia Steam Laundry."[34] During the same year, social reformer William Burns rebuked the growth of such businesses by suggesting that the laundress should organize and "repel with indignation . . . capital [which] has recently entered into competition with her." He declaimed "the extensive laundries which have recently been established" whose "hundred hungry arms are eagerly catching at every dirty shirt in the city."[35] Boston reformer Caroline Dall expressed a similar sentiment, suggesting the city should build a laundry for washerwomen to carry out their trade, providing them with supplies at wholesale cost so that they would have "a fair chance of competition" with laundries.[36]

Inventors and manufacturers of machinery also took notice of the new market. In October 1851, a Mr. J. T. King took out patent number 8,446 for a "washing apparatus" based on the same concentric cylinder principle as Strutt's earlier machine. King quickly put out a brochure and solicited testimonials.[37] By 1855, he offered thirteen different sizes of washing machines ranging from a hand-turned machine and machine C, "supplied with steam from a separate steam boiler and intended for small laundries where they have steam but no power," to model D #12, advertised as being suitable for large hotels, laundries, and public institutions. The larger machines came "complete with valves, driving gear, belting etc., ready to be connected with an engine."[38] The engine cost extra–presumably some places already had their own. King offered not just a better washing machine, but "everything for a complete laundry." His customers could buy either a single piece of machinery or an entire system including hydro-extractor, mangles, soap, and the engine to drive the machines.[39] Along with his first washing machine he also offered a variety of related products including a "magnetic washing liquid" (mild enough, he explained, to be drinkable in small doses).[40]

British firms offering commercial laundry machinery began appearing at the same time.[41] They tended to grow out of engineering concerns or foundries, offering a wide variety of products, technical advice, and custom-made machinery. One of the earliest and most successful British companies was Manlove and Alliot. Two Nottingham engineers originally founded the company to exploit a device called the "centrifugal extractor," reputed to have been invented in the early 1830s by a Swiss inventor named Seyrig. This machine had been intended for refining sugar, but the two engineering partners recognized its broader possibilities: the extractor could also be used to draw excess liquid out of wet fabric by means of centrifugal force. Alliot took out a British patent in his own name and the device was put into production. While it helped solve a critical problem in laundries, Manlove and Alliot also marketed it to bleachers, piece dyers, textile manufacturers and sugar refiners.[42]

The emergence of the industry led to the first comparison of the virtues of domestic and commercial methods and the first accusations of damage done by commercial methods. In her famous 1861 *Book of Household Management*, Mrs. Isabella Beeton briefly noted that many urban families were beginning to send their laundry out to be done by "professional laundresses and companies, who apply mechanical and chemical processes to the purpose." She pointed out, however, that "these processes . . . are supposed to injure the fabric of the linen"–better to retain a laundress and do the finer goods at home.[43]

Machine makers, advertisements, and complaining consumers provide evidence of the emergence of an industry but do not explain why the industry appeared when it did. The answer to that question lies in the cultural and material changes industrialization and urbanization brought to everyday life in the nineteenth century.

Making a Cleaner Society

Rising standards of cleanliness provided a crucial cultural context for a nascent laundry industry. New habits and practices emerged gradually over the course of the century. Looking backwards, these changes were stunning. Mountains of filth disappeared from the streets; overwhelming clouds of body odor that had hung over public gatherings became an experience of the past; vermin that had tortured rich and poor alike disappeared; and a host of skin-transmitted and waterborne diseases no longer made miserable or carried away whole sections of the population.

Although this new culture of cleanliness manifested itself in a multitude of ways, the habit of wearing clean clothes was one of its most visible and personal manifestations. During his old age, Englishman Francis Place remembered an eighteenth-century youth in which:

> The wives and daughters of journeymen tradesmen, and shopkeepers, either wore leather stays or what were called full boned stays. . . . These were never washed, although worn day by day for years. The wives and grown daughters of tradesmen and gentlemen even wore petticoats of

> camblet, lined with dyed linen, stuffed with wool or horsehair and quilted. These were worn day by day until they were rotten, and never were washed. A great change was produced by improvements in the manufacture of cotton goods.

From the vantage point of the 1820s, he noted that amongst the changes he had witnessed in his long life, "the increased cleanliness of the people is particularly striking."[44]

The rich of that vanishing age were perhaps a bit more scrupulous, aware that clean, white linen and a well-groomed person were increasingly taken to be signifiers of gentility. Writing in July 1747, Lord Chesterfield advised his son, "As you must attend to your manners, so you must not neglect your person; but take care to be very clean, well dressed, and genteel."[45] This paragon of gentility did not specify what "very clean" might mean. (Perhaps this nonspecificity has given the book its longevity as a guide to manners–each age applying its own interpretation.) A nascent American gentry was quick to take up the advice of Chesterfield and other standard setters.[46]

Still, admonitions to cleanliness did not have quite the same meaning to eighteenth-century moralists as to the Victorians and their successors. When John Wesley advised his followers that "cleanliness is, indeed, next to godliness" during the dirtier days of the eighteenth century, he had more modest aims in mind than the uses to which the phrase has since been put.[47] He hoped his parishioners would wash their faces and hands before coming to church on Sunday and would, perhaps, also change their shirts before entering the Lord's house. Despite a reputation for obsessive fastidiousness, Wesley probably did little more than this himself.[48] "Cleanliness," to the eighteenth-century reader, often meant something more akin to what we would call "neatness."

Frequent bathing was widely considered to be a threat to health, and the tub or shower was approached with great caution. Lingering beliefs dating from the late Middle Ages suggested that wetting the porous skin left one highly susceptible to disease.[49] Elizabeth Drinker, a late-eighteenth-century Philadelphian of scrupulous moral and social standing, went twenty-eight years without being "wett all over att once." In 1799, she approached that long-delayed washing with much trepidation.[50] Her change of heart was contemporaneous with the first wave of new hygienic habits in America.[51]

Particularly after the use of cotton underclothing became widespread, clothing was changed more often, customarily as a substitute for cleansing the skin. For those that could afford it, this often fell into a weekly pattern. Fresh linen was donned on Sunday, allowing one to face the Lord while wearing clean underwear.[52] This Sabbatarian habit persisted throughout the nineteenth century, as commercial laundries customarily returned finished goods to their customers on Saturday. The *linen cycle* became so deeply ingrained that even people who had no intention of attending church insisted on a Saturday return, putting pressure on laundries to work employees overtime to meet the deadline.[53]

Already changes were afoot in the connections educated people made between dirt and disease. The medical discourse on cleanliness gave even those urban dwellers unconcerned with social niceties reason to fear for their lives. Proponents of the newly introduced miasmic theory of disease believed that epidemics originated from noxious vapors–miasmas–rising from "the bowels of the earth," which could be detected by their foul smell. These ideas gained wide currency in both England and America by the middle of the eighteenth century, modifying older humoral theories that placed the cause of disease in imbalances within the body itself, not in the external environment. By the second half of the eighteenth century, the miasmic theory was firmly established, persisting long enough to compete with the discovery of germs at the end of the nineteenth century.[54]

The epidemic of waterborne diseases accompanying the growth of cities added an extra sense of urgency to understanding the relationship between dirt and disease. In the beginning of the nineteenth century, Britain became the world's first urbanized society. By midcentury, more British people lived in urban areas than in rural areas. Though the United States made rapid gains in the nineteenth century, only 13 percent were urban dwellers in 1850.[55] Still, U.S. cities grew rapidly, and residents of burgeoning New World metropolises gradually came to experience the same problems as British city folk.[56] As a reaction to the terrifying cholera epidemics of 1832 and 1848 as well as Edwin Chadwick's widely read parliamentary reports on the health of cities, the well-informed and well-placed gave voice and authority to a growing popular belief in the connection between squalor, foul odors, dirty water, and disease.[57] Americans also read Chadwick and recognized the problems that he identified as they pertained to the growing cities of the United States. The miasmic theory of disease and its successor, germ theory, moved across the Atlantic from the East along with the microorganisms that had inspired them.[58]

In a sense, the series of discoveries between the 1860s and 1880s linking microorganisms and disease just confirmed and gave justification for the already developing practices and beliefs. While true antisepsis would be impossible to achieve in ordinary life (despite the claims of some cleaning products), principles of antisepsis caused a revolution in medical procedures. Such ideas were given new urgency as society confronted a new invisible enemy: dirt could be lurking even where it could not be seen or smelled, bringing with it the threat of disease. Science had confirmed what popular morality had long asserted: one could never be clean enough.[59]

Although germ theory was the revolutionary discovery of the century, other discoveries were also highly influential in relation to cleanliness. A growing understanding of the processes of respiration and the renewal of the human body profoundly influenced many writers on hygiene. Part of this change during the eighteenth century was recognizing the connection between dirt and skin diseases. Filthy linen and a lack of bathing both aggravated many conditions such as impetigo and ringworm.

Having previously eschewed bathing and frequently changing clothes as unhealthy, inconvenient, and socially unnecessary, both Americans and the British gradually became voracious consumers of cleanliness during the nineteenth century. The impetus for this revolution came particularly from the middle class. Exposed to the dangers of unhygienic cities, wedded to commerce as a way of earning a living, and burdened with a pressing need to define and defend class boundaries, they had multiple motivations for adopting new standards–as well as the means, material and political, to impress those standards upon others. Cleanliness became both a health issue and a way to distinguish oneself from the laboring classes.

Despite the undeniable impact of scientific ideas, ordinary consumers clung more tightly to the symbolic definitions of cleanliness.[60] Many would-be laundry patrons still considered a stained and wrinkled shirt dirty even after it had been washed in hot water and soap, and that notion of an indelible lack of cleanliness contributed to prejudices against people (like laundresses) judged inherently "unclean" because of their race, class, or ethnicity. White middle-class Americans, in particular, believed that African Americans and certain groups of immigrants were fundamentally "dirty" in a way that no amount of scrubbing could cure. Therefore, people were concerned with how and by whom their clothes were washed. Cleanliness itself was gradually gendered, and came to be thought of as a female trait in the nineteenth century. Women were seen as the cleaner sex, better able to judge the clean from the unclean, better able to oversee the consumption of cleanliness.[61]

Although the British reformers adopted a preoccupation with the public health aspects of cleanliness before Americans, it is debatable whether the British as a people were actually any cleaner than Americans. The Atlantic did not serve as a cultural barrier.[62] Ideas and prejudices traveled easily between the privileged and commercial classes of both countries. English visitors to the United States commented on the filth in which American farmers and the rural poor lived, but travelers in both countries rarely remarked on the habits of members of their own class.[63] From the early nineteenth century, members of the English middle and upper classes also clamored for remediation of the habits and living conditions of English urban workers, suggesting that urban versus rural settings and class differences may have been more significant than transnational differences in setting standards.

Within a larger European context, both the English elite and the American upper classes carried the self-perception of being the cleanest of the Western world.[64] Writing for the American readers of *Harper's*, English author Harriet Martineau played to Anglo-American prejudices, using the example of a German aristocrat to parody older attitudes about washing: "When a German prince was told confidentially that he had dirty hands, he replied, with the liveliness of conscious triumph, 'Ach, do you call dat dirty? You should see my toes!'"[65]

Among the most prominent public symbols of these new-found values was the clean, white, starched and ironed shirt. Its careful preparation and presentation in public places preserved two important illusions for the middle class: it hid the fact that one worked for a living and concealed evidence of bodily functions such as perspiration.[66] Such symbols were particularly important in the growing anonymity of the urban environment where one was judged and identified mainly by appearance. Cleaned and pressed clothing signaled to strangers the finer gradations of social status and of urban *savoir-faire*.[67]

Clean linen came to be not only a prerequisite of hygienic living and a marker of class standing but also a measure of morality. New Yorker William Burns suggested that "when you come across a lady who delights in talking scandal, in telling the world of all the wickedness she knows there is in it, quietly mark it down in your memorandum book, that she is addicted to the wearing of dirty linen."[68] The moral construction of such notions of cleanliness was also closely tied with the function of cleanliness as a justification for social hierarchies. Scores of reformers echoed the anonymous author of a pamphlet on the necessity of baths and washhouses for the poor, who argued against the "crowded state of the habitations of the poor, which makes it impossible for them in many cases to cultivate those habits of cleanliness which are equally essential to the physical comfort and to the sound moral state of the populace."[69] Even the nineteenth-century "sanitarians," as they sometimes called themselves, wedded scientific discoveries to an older cultural linkage between cleanliness and morality to assess and reform the habits of "the great unwashed," as the lower orders were called in that most telling of Victorian phrases.[70] Science provided a rationalization for cultural beliefs and a language with which to talk about symbolic links between dirt and disorder.[71]

While reformers argued that it is difficult to be moral if one is physically dirty, others drew a different conclusion, arguing that if one is poor, one is probably dirty, and if one is dirty, one is probably not moral. Middle-class people continually confused the inability to get clean with the unwillingness to do so. While ignorance may have played some part in the poor's clinging to unhealthful habits (it was widely claimed that they believed a layer of dirt would keep them warm), the singular difficulties of getting and staying clean while doing manual labor and living without plumbing should not be underestimated.

Some of the most vivid examples of efforts to work out the relationship between cleanliness, morality, and class can be found in English religious tracts.[72] An 1823 work entitled *Cleanliness Is Next to Godliness* is an early example of the genre's ambivalence towards the connection between cleanliness and morality. The author, Lucy Cameron, tells the story of a woman so preoccupied with keeping her house and children spotless that she has no time to tutor them in religious matters. One of the children dies after having received deathbed instruction from the pious neighbor, Mrs. Read. The mother is later chided that she has put "godliness next to cleanliness," for, as Mrs. Read explains, "you have had the cleanliness, the order, the neatness of your family more at heart than the salvation of their souls."[73] One would be hard-pressed to find a tract writer half a century later making the same sharp distinctions between cleanliness and godliness.

In an 1852 sermon entitled *On Cleanliness,* G. B. Dickson laid out the reasons for attending to personal cleanliness, using a favorite Victorian moral device, the call to duty: "There are many considerations which ought to induce all to attend to the duties of cleanliness. It is a *personal* duty, that is to say, a duty which we owe to ourselves individually; it is a *social* duty, that is, a duty we owe to society; and it is a *religious* duty, that is, a duty which we owe to God"(1).[74]

Dickson then goes on to explain his thesis in greater detail. Cleanliness is in part a personal duty because one can only maintain self-respect and the sense of one's special place of having been made in God's image by maintaining the personal dignity that cleanliness brings. Moreover, "if we see a man with ill-arranged and slovenly dress, and dirty in his person, we are led irresistibly to the conclusion that he is a man of irregular habits, and of an ill-ordered mind; and that the indulgence of these habits not only betrays the inward disorder, but its influence is such as to increase and perpetuate it" (2).

Dickson's explanation of the social duty of cleanliness reveals the hegemonic function of such standards: "A person accustomed to cleanliness cannot be, and ought not to be, on terms of familiar intercourse with another who will not take the trouble to make himself clean. The separation of the classes of society . . . prevails to such a degree that little or no intercourse is a great existing evil . . . but it is a question worth considering how far this evil is increased and continued by the uncleanly habits of many" (3).

He goes on to charitably note that of course not every rich person is clean, nor is every poor one dirty (3).

He further informs the reader that "The great foundation of the social duty of cleanliness is, *that it is essential to the preservation of health*. . . . All medical men agree that frequent washing with water is necessary for the healthy action of the skin. And many diseases of the skin originate from attention not paid to this duty" (5). Finally, cleanliness is a religious duty because "No man, who is not careful to come clean in his person to the worship of God, can be prepared to worship God aright: for in his own person he is doing dishonour to Him in whose image he is made. . . . We must worship God with our bodies and our spirits" (6).

Dickson's reasons for attending to cleanliness were echoed throughout nineteenth-century British and American discussions on the subject. Taken together, these issues of health, social status, and morality provided a powerful driving force for new ideologies. Once endorsed by commercial culture, they became unstoppable as standards rapidly accelerated.

Like Dickson, the anonymous author of an 1847 British tract entitled *Wash [An Address on Cleanliness]* combines moral and practical reasons for cleanliness. The address begins with a very explicit description of the unwillingness of some people (particularly the poor) to wash themselves and their clothing. The text soon begins to mix the message of physical and spiritual cleansing until it is difficult to know which is being discussed: "God has given you a very plain and easy command. He has said 'wash.' You acknowledge Him to be God, Almighty, All-knowing. Then he must know whether you need washing—and you say, you don't. He, on the other hand, declares that you do."[75]

The conclusion preserves some of this ambiguity but suggests that both physical and spiritual reasons are being referred to: "Don't put off washing. People sometimes make their intention of washing on Saturday night an excuse for being dirty the rest of the week. So there are many who, when invited to 'wash' themselves from their iniquity in the blood of Christ, declare they mean to do so when they come to the last. There is no putting off to any of God's commands–no Saturday night. 'To-day' is the style to which they run. 'To-day' then 'wash.'"[76]

Another function of linking cleanliness with morality was to inculcate the habits of self-censorship and self-regulation on both counts. By these standards, the gossipy woman described by Burns would have known her underwear was dirty had her moral faculties been intact. The kind of anxiety engendered by the internalization of such standards made for almost compulsive attention to such matters among the conscientious.[77]

Attention to personal cleanliness was particularly important to those who, in one way or another, inhabited the fringes between social classes. Social climbing demanded a thorough knowledge of the niceties of proper grooming. The appearances of those with declining fortunes might reveal their inability to maintain standards: for those worried about the security of their class status, etiquette manuals provided advice for the reader about changing standards of cleanliness. The vehemence with which some of this advice is put is a testament to the degree to which cleanliness helped separate the rich from the poor. While American manuals skirted the issue of class, British examples tackled it head-on. *Routledge's Manual of Etiquette,* for example, declared that "in these days of public baths and universal progress, we trust that it will be unnecessary to do more than hint at the necessity of the most fastidious personal cleanliness . . . a soiled shirt, a dingy pocket-handkerchief, or a waistcoat that has been worn once too often are things to be scrupulously avoided by any man who is ambitious of preserving the exterior of a gentleman."[78]

Women were not exempt from such judgment. The Routledge manual repeated its advice almost verbatim in the section addressed to women, substituting soiled gloves and muslin dress as the offending articles of clothing.[79] An American manual, *The Lady's Guide to Perfect Gentility,* approached this issue of class and changing standards in a different manner: "What must we think of those genteel people who never use the bath, or only once or twice a year wash themselves all over, though they change their linen daily? Why, that, in plain English, they are nothing more or less than very filthy gentry; and you will find, if your olfactories are at all sensitive, whenever you happen to be near them, and their perspiration is a little excited by exercise, that they have a something about them which lavender water and bergamot do not entirely conceal."[80] In other words, if a person claimed to be one of the "better sort," that person had better wash accordingly.

In the more ambiguous context of American social ranking, a veneer of cleanliness aided upward mobility. A standard device in almost all of Horatio Alger's stories is the transformation wrought by a new suit of clothes. Nowhere is this device more apparent than in Alger's first novel, *Ragged Dick* (1866). In a chapter entitled "Dick's New Suit," the protagonist's benefactor, Mr. Whitney, provides a suit and admonishes Dick to wash before putting it on because "Clean clothes and a dirty skin don't go very well together."[81] Thereafter, a series of adventures ensue as old acquaintances fail to recognize Dick in his new finery and new acquaintances (including a dealer in stocks) mistake him for a member of the privileged classes.

By the end of the nineteenth century, middle-class values had made a notable impact on working-class notions of the relationship between cleanliness, clothing, and community standards. "Cleanliness is our greatest class-symbol," claimed Stephen Reynolds in his 1908 fictionalized account of life with a Devon family, *A Poor Man's House.* "At the bottom of our social ladder is a dirty shirt; at the top is fixed not laurels, but a tub! The bathroom is the inmost, the strongest fortress of our English Snobbery."[82] In his memoir of youth in the Manchester slum Salford, Robert Roberts notes that clean, unragged clothes were one of the most important maintainers of status. "Tradesmen took pride in wearing stiff collars even in the 'workshop' (a 'factory' was where the unskilled laboured). A child's weekly chore was to have this neckwear at the laundry and returned in time for father to choose the best out of a frayed collection and go off, 'dressed to the nines' on his Saturday evening booze."[83] Bedding and underclothes in good condition were also important, since they would be on view when hung out on washing day. "A poverty-stricken display" notes Roberts, "could do one much social damage."[84]

Servants must have been particularly aware of the ironies involved in the class lines that cleanliness helped define. They worked on the front lines of keeping the middle class clean while enduring limited opportunities for themselves. As public rhetoric announced the importance of cleanliness for people at all levels of society, private debates centered around issues of whether to let the servants use the tub and which ethnic group was naturally cleaner. Employers were warned that "If a girl is uncleanly in her person, there is every reason to believe she will be slatternly in her work."[85] Still, employers had an obligation to keep up their own standards as an example to servants and children.

Science and morality, fear of disease and desire for social status, proved a potent combination. Hard-earned wages invested in that ephemeral good–cleanliness–was money well spent, buying both respectability and health. Increasingly, it began to seem that one could never be too clean.

Urban Infrastructure

By the mid-nineteenth century, fastidious urban dwellers desiring crisp collars and snowy table linen had several laundering options: do the work themselves, find a washerwoman, or patronize a laundry. The sooty air and foul water of growing industrial cities created an environment that gradually gave the advantage to mechanized and highly organized laundries over other commercial alternatives, since washing at home had become increasingly impractical. Washerwomen contended with arduous trips to the standpipe for water, fought to keep soot out of the washtub, and hung wash from the ceilings of tenement rooms–a poor substitute for the hedges and drying fields of a more rural age. Meanwhile, laundry owners sealed their windows to keep soot away, paid for piped water, and directed their employees to hang the washing in a drying closet.

The key to both sanitary reform and the growth of laundries in an urban context was a steady, reliable, clean supply of water. Before industrialization, city dwellers had suffered few problems with pollution, except in a few huge metropolises such as London. Obtaining water was a private matter. Cities sometimes provided public wells, but little more. As the quality of water began to decay in the late eighteenth century, private companies most often took up the slack. Most were quite small, supplying water from springs to a few thousand stockholders in towns and villages.[86] Although London was one of the earliest cities to have an extensive water system made available through private companies, a welter of commissions, boards, companies, and individual entrepreneurs who cleaned the streets and supplied water from water wagons all jealously held off centralization until 1899.[87]

By midcentury, a number of large cities in both America and Britain had embarked upon municipal water projects, spurred on by new hygienic ideas and the fear of waterborne epidemics.[88] Despite talk about the terrifying conditions in slums, the poor benefitted least from such enterprises. Companies laid pipes where convenient and profitable, generally heading in the direction of wealthier areas. Even after the great municipal water projects, many still depended on streams, wells, or on the common standpipe or pump. In 1883, George Sims, an observer of the London poor, noted that "in many houses more water comes through the roof than through a pipe, and a tub or butt in the backyard about half full of a black, foul smelling liquid, supplies some dozens of families with the water they drink and the water they wash in as well."[89]

In the competition between washerwomen and commercial laundries, inequitable water distribution meant that the women often walked substantial distances to bring home buckets of water of dubious quality.[90] Given the degree of pollution in some urban water supplies, washing could be a futile measure, linen sometimes emerging more gray than before it had been submersed. In contrast, laundries could afford to have cleaner water piped in, and, by the end of the century, large laundry businesses sometimes installed elaborate systems to soften and purify it.[91] Laundry workers spent their days involved in processes that used thousands of gallons of water, and yet some went home to households without running water and in which hot water out of the tap was virtually unheard of.

Air pollution also figured into the challenges that urban laundresses had to face. An 1877 servants' manual, *The Laundry Maid*, noted that "it is necessary for the London laundry woman, whatever her station, to remember that clothes become dirtier much sooner, or much dirtier with the same amount of wear, in London than in the country." Laundresses were advised that they should "dry off the clothes as quickly as possible, else they will certainly become partially dirtied again before they are taken from the line."[92] A small verse, probably composed by a laundry owner, described the problem in humorous terms:

> Sing a song of Monday, weather nice and fine
> Four and twenty washes hanging on the line;
> Soft coal's what they're burning, woe in many a flat
> Every single garment's blacker than your hat.[93]

In short, the commercial establishments that were beginning to be introduced in the second half of the nineteenth century were simply better equipped than the home to combat the rising tide of urban filth. Laundries sealed their windows against the "blacks" (as particles of soot were called in England) and dried the clothes inside.[94]

The nascent laundry industry benefitted directly from the cultural and physical changes that accompanied industrialization and urbanization. In its heyday, the industry successfully catered to shifting tastes and beliefs about cleanliness, merging public attitudes about mechanized processes, technical expertise, and hygiene. Laundries touted their services as the modern, hygienic version of more traditional processes. They offered the convenience of assistance in a household process without the difficulties of managing servants; foul mountains of household linen disappeared behind factory doors to emerge fresh, clean, and ironed.

Laundries served a wide variety of clients, all characteristic of new urban lifestyles and aspirations. The "bachelor bundle" was a mainstay of some laundries, serving the needs of an army of clerks and businessmen whose white collars asserted their rising status. Many middle-class households sent out everything but delicate goods and dresses made of unwashable fabrics. A growing number of travelers on steamships and Pullman cars generated mountains of soiled towels and sheets–accoutrements that were becoming an expected part of civilized travel. By the 1880s, Pullman boasted clean sheets each night on its American sleeping cars.[95] Other nineteenth-century institutions striving to replicate the functions of a household on an institutional scale either created steam laundries as part of their in-house facilities or contracted out for laundry work. Hotels, schools, and lunatic asylums confronted the problem of turning out huge amounts of clean linen on a daily basis.[96]

The combination of new fashions and new methods of manufacture meant that, in addition to flatwork and undergarments, the public uniform of the Victorian male was also well suited to the factory methods of the laundry. The nineteenth century has been described as an era of remarkable conformity in male dress. Men of commercial aspirations emerged from their houses and flats looking as alike as the goods stamped out in the growing number of factories that dotted the landscape. Underneath their black suits, they wore relatively simple white shirts. The introduction of detachable collars and cuffs in the 1830s allowed men to change the most visible parts of their linen without putting on an entirely fresh shirt–thus cutting down on laundry bills.[97]

Unlike the male frippery of the eighteenth century, with its ornate fabrics and complex construction, these shirts, cuffs, and collars lent themselves relatively easily to relaundering. Late in the century, the shirtwaist offered the same advantages to women. Likewise, the flood of cheap cotton cloth spilling out of new mills replaced the wool and linen of earlier centuries which had been more difficult to launder. Black suits, of the appropriate hue to hide the layers of soot from coal smoke, were less often washed or sent to be re-dyed.[98]

Mechanized laundry appealed to consumers for reasons other than just how clean the clothes could be made. Middle-class consumers found the idea of having their clothes washed in the possibly squalid and disordered atmosphere of the washerwoman's tenement distasteful. Bringing a laundress in disrupted the household, filling the kitchen with steam and wet laundry. Washerwomen were also thought to be a corrupting influence on other household servants, who might linger in the washhouse or kitchen to gossip and be inspired to insubordination by the independent ways of the laundress.

Though women of the same social order as the private washerwomen worked in power laundries, emphasis was put on the ma-

There Are Three.

Society is divided into three classes—the men who wear clean linen, the men who wear soiled linen, and the men who wear no linen at all. Laundries are divided into three classes—those which do good work, those which do poor work, those which do no work at all.

We do good work—lots of it. It would do us no good to say this if it were not so. Your money will buy more cleanliness here than anywhere else.

Fig. 4. This illustration from an 1899 advertising pattern book neatly summarizes the class and moral connotations of shirts and collars in Victorian Britain and the United States. From Charles Austin Bates, *The Laundry Book* (New York, 1899) n.p. Courtesy of Hagley Museum and Library.

chines, not the workers—who were portrayed as not in control of the process but rather there to function as an adjunct to the machines. Under the watchful eye of a manager, these workers were held to his or her better understanding of middle-class standards for cleanliness. Managers were further seen to have replaced the washerwoman's self-determinant workstyle by enforcing work pace and decorum among the workers. The names of laundries furthered this perception of the superiority of highly mechanized processes and middle-class control in achieving cleanliness: most evoked either whiteness or cleanliness with names like "Sanitary" or "Pearl."

In 1920, American writer Anzia Yezierska published a short story entitled "Soap and Water" about a woman from a poor family who is working her way through college as a laundry worker. The dean of women withholds her diploma because, as her middle-class instructor informs the student, although her work is excellent, the dean cannot recommend her as a teacher on account of her personal grooming. "Soap and water are cheap. Any one can be clean," explains the dean.[99]

The phrase echoes through the rest of the story as the narrator explains that her flat has no bathtub and that she has no time for the kind of personal grooming the dean demands. As she stands at her ironing board she meditates on the irony of the situation: "I, soaking in the foul vapors of the steaming laundry, I, with my dirty, tired hands, I am ironing the clean, immaculate shirtwaists of a clean, immaculate society. I, the unclean one, am actually fashioning the pedestal of their cleanliness, from which they reach down, hoping to lift me to the height that I have created for them."[100]

✕

How and why does the history of this industry begin? Fred De Armond and other laundrymen answered: at the moment the first forms of technology that eventually distinguish power laundries first appear. That answer provides a simple solution to a complex question. Unfortunately, the technological logic of these stories does not hold. As this chapter has shown, machinery with the potential for commercial use was available long before it was taken advantage of. Patents, likewise, are no marker of commercial employment.

Alternatively, the history of an industry might be said to begin when both the technological means and the cultural need make such enterprises economically viable. If so, the beginnings of this industry were located not in a single moment of inspiration but in a long series of decisions made by consumers, entrepreneurs, and creators of technology over the course of decades. Entrepreneurs chose to invest in laundries as businesses. Consumers elected to patronize laundries rather than do the wash themselves or engage a washerwoman. Both were motivated by a complex set of factors, including new standards of cleanliness and a degraded urban environment that made traditional methods less and less desirable.

2

The Business of Laundry

By the 1880s, more than a few great men, like William Strutt, found themselves thinking about laundry. A growing number chose to invest their capital and time in the laundry business. The result was the emergence of what contemporaries called the *laundry industry:* a phenomenon characterized not just by individual businesses springing up in towns and neighborhoods but also by the appearance of journals and associations, suppliers and engineers, and an international trade in laundry equipment.

On February 12, 1884, one such group of men announced the formation of a joint-stock company for the purposes of opening the first steam laundry in Newcastle-upon-Tyne. Most of the stockholders were connected to the coal mining and metal trades that fueled the economy of that grey, northern port city. While some left no record as to why they would divert their capital into this new business, at least two investors had well-defined motives. James Joicey (later Lord Joicey, M.P.), one of the richest mine owners in England's North Country, sought a means to set up his ward, Walter Allan, in business. Through Joicey's agency, Allan soon became the company's first manager and a minor stockholder. A smaller amount of stock went to James Armstrong, a local ironmonger who had just entered the business of importing American laundry machinery. Armstrong stood to profit not only from the laundry's receipts but also from positive publicity for the machinery he provided for the laundry.[1]

The stockholders named their new enterprise *The American Steam Laundry Company*, to advertise the origins of its machinery. They secured a ten-year lease on a disused skating rink underneath a veterinary stable (later remembered as having filled the laundry with a very strong smell of horse). A staff was hired, customers solicited, and a Mr. Ware put in charge until the machinery was running well enough to be handed over to the inexperienced Mr. Allan.[2]

The residents of Newcastle soon bombarded the board with complaints of "bad washing," "unpunctuality in delivering goods," and "mistakes in delivery." At a December board of director's meeting, Armstrong confronted Allan, who had been in charge for several months, with the bad news. Allan responded lamely, blaming the workers and the machinery (earning the enmity of Armstrong, whose boiler and mangles he derided). The situation had not improved by February, when an audit showed a loss of nearly 300 pounds. The board apparently decided it valued profit above the bonds of nepotism, voting to replace Allan with a young Welshman named William Bird as supervisor of the laundry's day-to-day workings.[3]

Bird entered the business not as a rich man's ward but rather an ironmonger and practical engineer, newly released from his apprenticeship. He had originally been hired to help install machinery, but once in charge, Bird proved himself a more capable manager than his predecessor, and gradually the laundry began to prosper.[4] After a year of secretarial duties, Allan quit the laundry business altogether. Armstrong also fell from grace owing to his unwillingness to replace defective machinery, including an unreliable boiler. Initially only removed from the Board, Armstrong and his representatives were eventually forbidden from even entering the laundry.[5]

In Armstrong's case, initial failure was no predictor of eventual outcome. He left Newcastle for London and later made a name for himself as the premier importer of American laundry equipment to Britain.[6] The American Steam Laundry Company was renamed *Bird's Laundry* in 1911 and prospered under its new title. It became a Newcastle institution, buying up other regional laundries as they failed and providing services to local elites and coal miners alike.[7]

The longevity of Bird's Laundry and the survival of its records make it unusual. However, many of the details of the firm's origins show the business of laundry as it emerged in the late nineteenth century. Joicey and his fellow board members were among the increasing number of businessmen eager to invest in washing clothes. During the 1860s, the first laundries began to appear on the London stock exchange, and by 1879 the public could invest in syndicated firms.[8]

The presence of James Armstrong in this story reveals not only the importance of machine dealers in the success or failure of laundries but also the quite early existence of a transatlantic business culture. Although British companies already manufactured laundry machinery, very little of it was exported. Technology flowed primarily from the United States to Great Britain. By the turn of the century, a number of American laundry machinery companies advertised sales offices in London. However, with the exception of a few laundry chemicals, British laundry technology does not seem to have been marketed in the United States.[9]

Allan's inability to get the laundry to function properly exemplifies a more widespread phenomenon. In his naiveté and eventual failure, Allan had plenty of company. What seemed like a simple enterprise—low-tech, straightforward, and serving a proven need—often turned out to be much more complicated. The technology was not idiot-proof, and it was forever changing; workers were not as tractable and dependable as their employers had hoped; and competition loomed on all sides, not only from Chinese laundries and washerwomen but also from the customers who might choose to do the work themselves.

The American Steam Laundry Company also shared another characteristic with the other laundries that appeared in the following decades: women constituted most of the workforce, but the business was owned and run by men. Joicey apparently thought it perfectly fitting to discharge his duties as guardian by putting his ward in a laundry rather than the coal or metal trades. Likewise, Bird traded in the grit and male camaraderie of the ironmonger's shop where he had spent his apprenticeship for the smell of soap and the supervision of women. Both its machinery and its public image of male authority set this kind of laundry apart from the independent women who comprised its competition. One American laundryman was blunt about the shift: "With all due respect to the fair sex, real progress only began after business men began to see the potentialities in the business and to invest both their money and their brains therein."[10]

Scope and Scale

In later years, aging laundrymen recalled laundries such as Bird's as industry pioneers. While not the first to offer such services or to call themselves "steam laundries," these laundries were frequently the first in their provincial towns, the first to survive into the twentieth century, and the first to identify themselves as part of an industry.[11]

Because nineteenth-century American census takers ignored the industry entirely and their British counterparts did not distinguish between steam and hand laundries, demographic information about the industry during the 1880s and early 1890s is scarce. In 1909, the first census of power laundries in the United States counted 5,186 establishments. Census takers ignored hand laundries, limiting their enumeration to facilities that used power-driven machinery. They also restricted their count to commercial establishments, leaving out the scores of laundries in hospitals, penitentiaries, and charitable establishments (some of which made a goodly profit off the labor of their inmates).[12]

The British census of 1900 counted laundries registered under the Factory Acts. This included a far wider range of establishments than the American census. Bitter struggles over broadening the scope of the law led to the inclusion of all types of laundries with more than five employees, including hand laundries and those in charitable institutions. This count revealed 7,021 registered laundries in England and Wales. A special 1904 census attempted to distinguish between power and hand laundries for the first time. The results were handicapped by a low rate of returns, but census takers eventually estimated approximately 2,254 British laundries using power machinery.[13]

Census takers also took note that laundries were a substantial source of employment, especially for women. The 1909 U.S. census counted 124,214 persons employed at wage labor in laundries (excluding salaried employees such as clerks and managers), 71.2 percent (or about 90,000) of whom were women. Previous to the recategorization of laundry work as manufacturing, it was considered a service occupation. Before this distinction was made, census categories lumped together those independent workers that took in laundry or did it in the customer's home with workers in commercial laundries. The 1907 U.S. census report on women at work, using the figures collected in 1900, counted 328,935 women working either in factory laundries or at home and an additional 25,638 Chinese men employed primarily in "Chinese laundries."[14]

The British census of 1900 revealed that the 7,021 registered laundries in England and Wales employed 196,141 females. The 1904 survey by the factory inspectorate found 59,104 employed in power laundries, 58 percent of which were females. British and American census returns also reported nearly the same proportion of laundry workers to the overall population for both countries.[15]

Both British and American counts also showed regional differences. London was the undisputed center for British laundry work, with whole neighborhoods (such as Acton) given over to the business of laundry. Laundries were also common in port cities such as Liverpool, Portsmouth, and Newcastle. Meanwhile, American laundries clustered in the cities of the Northern states. As reported in 1910, New York ranked first with 508 laundries, 126 in New York City alone. Illinois was second with 448, and California, although twelfth in population, came in third with 321–perhaps reflecting the difficulties of obtaining servants in the far west.[16]

This concentration in British and American cities shared some causes. New York, Chicago, Los Angeles, and the San Francisco Bay area were all tourist centers and terminals for rail or ship travel. In areas of the Southern United States, because of the persistence of washerwomen, large laundries were almost entirely absent. Outside of the larger Northern cities, cheap labor and the tradition of bringing a washerwoman into one's house stilled the growth of commercial laundries until the labor drain of World War II. Just before the First World War, Louisiana, for example, had just twenty-five and Florida twenty-eight. Only several Western states had fewer: Nevada had five and Arizona (with plenty of fresh air and sunshine) eleven. Other differences not always revealed by statistics also emerge. New York City had a preponderance of small shop-front hand laundries, while Chicago businesses were notable for their use of machinery and the use of large plants.[17]

The term *laundry* hardly describes the diversity of the establishments that fell under this title. With expansion and mechanization came both increasing diversity and contrasts in scale. Most of the earliest power laundries mainly washed men's shirts and collars and took in industrial work from hotels and ships. As machinery became more efficient and reliable and competition became stiffer, laundry owners invented new ways to take advantage of technological possibilities, offering to do the whole family's wash; to provide cheap, partial services for those with limited means; and to launder towels and uniforms from sources as disparate as the barbershop and the executive washroom. The W. H. Garlock and Company Laundry and Shirt Factory, for instance, warranted a half-page spread in *The Industries of Cleveland* in 1888. The text predicted that "ere many years 'wash-day' with its miseries will have forsaken the American household forever."[18] By 1900, the company had abandoned shirt manufacture for full-time laundry work and its owner had gained enough civic prominence to have his photograph in *Men of Ohio in Nineteen Hundred*.[19] The aforementioned Bird's Laundry began by specializing in "starch work"–shirts, collars, and cuffs–but quickly branched out into what became known as "domestic laundry" or "family wash."[20]

Large, heavily mechanized laundries like Bird's were the exception rather than the rule during the 1880s and into the 1890s. By the turn of the century, plants ranged in size from shop-front establishments with less than five employees to huge factory laundries that might take up an entire city block, employ more than two hundred people, and process fifty thousand pieces a day. In the same city, some laundries might employ only one boiling copper and a stove-heated iron while others utilized ranks of massive belt-driven washing machines and mangles. Some took in only family washing while others washed the mountains of towels and sheets unloaded from Pullman cars or steamships. Meanwhile, hotels, hospitals, and schools all bought machinery and built their own laundries.

Observers were quick to note the rapidly growing number of mechanized laundries, but they were less eager to point out that, even after the turn of the century, large numbers of "hand" or "cottage" laundries using little or no machinery continued to exist. Moreover, in practice, the line between the two was sometimes ambiguous. "Hand" laundries, which often claimed to use no machinery at all, gave up their business to steam laundries only gradually. H. Llewellyn Smith, in his 1930 *New Survey of London Life and Labour*, estimated that in 1902, 29 percent of London laundries were mechanized. By 1912, that percentage had increased to 46. In 1930, hand laundries still constituted 30 percent of laundry businesses in London.[21] However, as steam laundry owners liked to point out, that distinction became increasingly meaningless. Many hand laundries gradually developed a peculiar relationship with their larger mechanized counterparts. Washing was subcontracted to mechanized laundries and returned to the hand laundries for starching and ironing.[22] Small neighborhood shops in London suburbs often did no real laundry work at all but instead acted as receiving depots for centralized factory laundries.[23]

Consonant with the rhetoric of industrial decline, British trade journals portrayed American laundries as larger and more mechanized than their British counterparts. They trotted out the usual evidence in support of this claim, postulating that Americans were motivated by the higher cost of labor in the United States; that they fell in love with novelty while their British cousins stubbornly clung to the past. Others maintained that the extension of the Factory Acts to laundries from 1895 to 1907 had pushed British laundry owners to mechanize. Effective regulation was not a problem most American laundry owners had to deal with in this period. Without better statistical evidence, assessing claims of differential rates of mechanization remains difficult. For my study, it is more important to recognize that although British and American laundry owners liked to emphasize their differences, the largest laundries on either side of the Atlantic resembled one another much more than they did their local, small-scale competitors.

Statistics about ownership also reveal the industry's complex structure. The 1909 American census figures show almost equal divisions of ownership between corporations (22%) and partnerships or firms

(24.3%). Individuals owned the remaining 53.7 percent.[24] Though the census did not correlate ownership with size of laundry (as measured in either capitalization or number of workers), individual ownership was probably concentrated in smaller, "family" laundries, whose proprietors had few assets to protect through incorporation and little extra money to pay for such legal services. Though the British census did not break down ownership, similar patterns can be ascertained from the trade journals. For instance, in 1899, the *Laundry Journal* noted a "steady growth of steam laundries owned by a limited company or syndicate, with thousands of pounds of capital, extensive premises, scores of hands, and the latest machinery."[25]

While there is not enough data to suggest a demographic profile for laundry owners, more scattered evidence suggests something of the background and culture of these people. The owners of most small laundries were white men neither well-educated nor well-off. Perhaps because of the identification of laundry work with domesticity, laundry ownership also attracted some women. With the exception of a few rare widows and daughters of male owners, most women laundry owners seem only to have run the smallest of laundries, contributing their own labor to the survival of their business. The British trade literature also singled out well-to-do laundry owners–"laundry hobbyists" or "cottage launderers" as they were sometimes disparagingly called. Advertisement in the back of trade journals further suggests that some middle-class women looked upon laundry ownership as a source of income.[26]

Many of the most successful laundry owners rose from the ranks of male laundry employees. Former deliverymen, managers, and inside workers knew how to navigate the industry's pitfalls. In 1958, the American Institute of Laundering published a retrospective booklet "Leadership through the Years, 1883–1958" for their annual banquet. It included reminiscences of a number of successful laundrymen. William Frew Long, a former president of the association, recounted that he had started in the laundry business in 1897 as the driver of a delivery wagon after his doctor had advised him to give up his bookkeeping job.[27] Another owner, Bert Thomas Geltz, explained that he had stumbled into the business after seeking a job as a shirt ironer at a hand laundry. "I knew almost nothing about it but I wound up buying the place for $100 down. The seller was to teach me how to iron shirts. He got drunk and never showed up. I was on my own and have been ever since." Other entries included an ex-lawyer and a graduate of Georgia Tech who both joined the trade in the same year as Long. The laundryman from Georgia Tech had "selected the laundry business because it was on a weekly cash basis" and because he had "the engineering qualifications" that enabled him to "look after the equipment."[28] A few laundry owners created veritable empires of laundries. Henry W. Stoer of Cleveland, Ohio, was not only president and principal stockholder of two Cleveland laundries, but he also owned interests in a number of New York and Chicago laundries.[29]

Other laundries evolved out of closely related businesses. The Millbay Laundry in Plymouth, England, began in 1890 as a tailoring and cleaning shop in Plymouth, in the south of England. The proprietor, Harold Roberts, had apprenticed as a tailor and made his living outfitting Irish immigrants to the United States, but he abandoned his family business for the laundry trade. In its early years, he ran the business out of the basement of the tenement where his family lived, employing his wife and young children to help with the work.[30]

A Community of Laundrymen

During the 1880s, the owners and managers (and, more rarely, the employees) of laundries began to describe themselves not just in relation to individual businesses but as part of the "laundry industry." By *industry* they meant not just an economic entity but a kind of imagined community that extended beyond the walls of their individual laundries or even the edges of the cities within which they lived and worked. This community was created through trade associations, journals, and annual conventions. They also gave themselves a name; they shared the commonality of being "laundrymen."[31]

There was little originality in the way laundrymen went about creating this community. They modeled their journals and associations after the trade journals and trade associations of other industries becoming common on both sides of the Atlantic during the same period. American laundrymen's efforts slightly predated their British counterparts, but only by a few years. From the beginning, people and ideas in the industry slipped easily back and forth between the two countries.

The voice of the laundrymen was their trade journals. The first British periodical of the industry, the *Laundry Journal*, began in 1885. Its competitor, the *National Laundry Record*, appeared six years later. The American publication, the *National Laundry Journal*, dates to 1884. This journal's competitor, the *American Laundry Journal* appeared in 1896.[32] Almost from the beginning, British and American journals appeared remarkably alike in their format and the content of their articles. This similarity sprang in part from editors' tendencies to liberally reprint articles, as well as the common prevalence of advertisements from companies such as the Troy Laundry Machinery Company, which marketed equipment in both countries. Some laundrymen also took busmen's holidays, reporting on the laundries of their transatlantic fellows and sharing in the conviviality of conventions far from home.

These editors manifested a fondness for using cultural stereotypes and odd bits of local color to highlight national differences. The British editor of the *Laundry Journal*, for example, found the American problem of competition from Chinese laundries fascinating. Puzzling over a particularly colloquial passage, another English editor reprinted it for his readers: "Plute squaws have taken the places of the Chinamen bounced from Truckee, and do the laundry work of the community," and he solicited help to interpret *Plute squaws* and *bounced from Truckee* ("Plute squaws" were female members of the Paiute tribe and Truckee is the name of a town in the Sierra Nevada Mountains).[33]

Beneath these bits of whimsy lay more subtle differences in tone. Between 1885 and 1920, the *Laundry Journal* dominated the British field. Unlike its American counterparts, this publication claimed to speak not just for laundrymen but for the "whole trade" (though the editor's idea of inclusivity was to describe company parties, births, and marriages). British journals also tended to have a more self-critical tone compared to the unabashed boosterism of their American counterparts.

The most important manifestations of laundrymen's efforts to build community, however, were the laundry trade associations. According to industry sources, Massachusetts laundrymen founded the first of these organizations in 1880. Three years later, laundry owners from around the country met at the Hotel Florence in Pullman, Illinois, to form the Laundrymen's National Association (LNA).[34] In 1886, the *Laundry Journal*, newly founded, carried an editorial pointing out the need for similar organizations in Great Britain. "It is a little curious and not a little unsatisfactory, that almost alone among the numerous important industries which are carried on in London, the Laundry Trade should have no society or association for the protection of its interests." The author of the article opined that an association was needed to fix prices, to deal with customers who made fraudulent claims and employees who stole, and to share technical information.[35]

Soon thereafter, London laundry owners met in a Holborn restaurant to discuss the matter. The two Americans in attendance told their British hosts they had had significant roles in the founding of the LNA. Because of their experience, they apparently felt free to dispense advice. A Mr. Felton told the gathering that "great difficulties had been met with at the outset" of the LNA's founding but that "eventually they had been surmounted. The chief of these had been the spirit of antagonism in the trade which was so great that, if one laundryman met another in the streets he would cross the road to avoid him."[36]

As it turned out, the LNA proved a durable and stable organization, while a succession of British organizations failed to attract and keep an adequate membership. Relationships between members of the British laundry community remained far more divisive than those between their American counterparts. Nevertheless, trade association organizers in both countries fought a steady battle to attract and retain members. Many laundry owners were reticent about joining a trade association; given their narrow profit margins and long workdays, they needed to be convinced that the time demands and expenses of membership were worthwhile. Associations, meanwhile, continually complained about laundry owners who reaped the benefits of the association's work without paying dues.

Trade associations tried to lure members with offers of both material and social benefits. For laundry owners who were interested in the long-term stability of their business, trade associations offered various means of making competition with other laundries more predictable: price fixing, the sharing of technological information, and the setting of industry-wide standards of service.

In the early years of these associations, price fixing was the most common means of stabilization. In the United States, the antitrust stance of the courts made explicit price fixing illegal after the turn of the century, although local groups sometimes fixed prices anyway. In the 1898 Illinois Supreme Court case *Doremus v. Hennessy*, Miss Mary G. Hennessy won a judgment of $6,000 against representatives of the Chicago Laundrymen's Association. These gentlemen had worked to force Miss Hennessy out of the laundry business because her laundry was undercutting the Association's price structure, but it seems they had picked the wrong laundress. Miss Hennessy had opened the laundry in order to pay for her legal education. She sued the association and won.[37]

Parliament never rigorously regulated price fixing in Britain during the period encompassed by this study.[38] However, not all laundry owners saw the potential to participate in monopolistic practices as an incentive to join a trade association. This absence of regulation of price fixing in Britain may, in fact, have contributed to the greater instability of British associations, as members joined or quit depending on market conditions. Jobbers and cut-rate laundries were particularly likely to be outside the fold because they survived by offering extremely low prices to customers or by subcontracting work from higher-priced laundries.

During most of the year, these early trade associations existed only in the actions of their leaders, in articles for trade journals, and as a mailing address (and, later, a phone number). For a few evenings or weeks of the year, conventions and dinners became a means not only to share information but also to create a community that existed in the same time and space for a few hours or a few days. Industry leaders believed that conventions were critical in cementing bonds between laundrymen. In large numbers, the laundry owners descended upon an appointed hotel for endless readings of papers, long dinners, and various entertainments. Conventions seem to have been an American practice that was adopted in Britain after the turn of the century.[39]

In an age profoundly concerned with the detrimental effects of big business and monopolistic practices on workers and small business owners, trade associations were careful to portray themselves as democratic and equally welcoming to everyone in the laundry community. In the editorial column of the *American Laundry Journal*, the editor extended a special invitation to smaller operators to attend the laundrymen's 1911 national convention: "It will make no difference how young or small your plant is, or whether you come in your working clothes or in your Sunday clothes, the reception committee will see that you are properly taken care of, and given all the attention and courtesy possible to be given."[40]

Laundrymen made two exceptions to this blanket welcome, one implicit, one explicit. Women could not be "laundrymen," although it was accepted that they might manage or own a laundry. In this process so identified with femininity and domesticity, women were the "other" inside the community. In the beginning of the industry, these boundaries were so rigid that not even laundry owners' wives attended conventions. By the start of the twentieth century, barriers had begun to break down. In practice, the British associations were quicker to admit women to the real business of the conventions and to positions of authority. This was a consequence of the importance of women as managers in British laundries (most American laundry managers, meanwhile, were men).

Nonwhites, male and female, were the explicit "other." In the United States, the LNA and some local laundry associations explicitly defined one of their roles as that of helping white laundry owners to more effectively compete against Chinese laundries. The minuscule number of African American laundry owners who appear in the census are invisible in the rhetoric and the membership of trade associations. This may be because they did not compete with white laundry owners but instead served the black community.[41]

Inclusion or exclusion in the laundry owners' community had practical consequences. Business friendships gave men access to capital. Prominent men could vouch for a smaller entrepreneur's credit worthiness or could simply provide an outright loan. These connections also granted access to information, a particularly important benefit for the inexperienced laundry owner or manager who was still trying to get control of a business. The only way to learn how to effectively run a laundry was practical experience and advice from others in the industry. The would-be entrepreneur needed a network of sources for counsel: laundry owners, supplymen, and others who would be willing to share information. As government regulation of laundries intensified after the turn of the century, trade associations also sometimes provided legal representation for laundry owners. More often, they represented the interests of the trade as a whole by lobbying lawmakers.[42]

The growth of these forms of association also had cultural consequences. Learning how to be a laundryman encompassed more than mastering the technical details of the industry. Conventions and dinners provided opportunities to create and inculcate ideas about masculinity and masculine culture in the business. The behavior of laundrymen on these occasions supports some historians' assertions that trade associations were one of the forums that began to replace fraternal organizations in late-nineteenth-century American culture.[43] Laundrymen traveling in their own Pullman car to the LNA meetings in Boston or Chicago or Buffalo shared inside jokes and learned about norms of behavior from their fellow travelers. Convention speeches and activities conveyed more than technical information. It was noted that British male conventioneers and diners showed more restraint at these gatherings than their American counterparts. Cultural stereotypes aside, part of the difference might have lain in the large female presence amongst the managerial ranks of large and small British laundries.

Likewise, trade journals also dedicated column space to discussions and exemplars of appropriate manly behavior. Because of the strong feminine connotation of traditional laundry work, this discourse often took the form of distinguishing what laundrymen did from what laundrywomen did.[44]

In 1924, the American journal *Laundry Age* published an advertisement that, although printed a few decades after the founding of the industry, captures the gendered and spatially distanced character of this emerging community. "Tell These Men About It," the headline reads. "Several thousand laundryowners read each issue of *Laundry Age*. If these men lived in one city and constituted its population, you would want to locate in that city provided you wanted to buy or sell or transact other business with them." The classified section of *Laundry Age*, the ad suggests, is the national equivalent of the classified section of that hypothetical city's local newspaper. The accompanying picture portrays those men sitting around a table together reading trade papers.[45]

Laundry Trouble

If many small laundry owners eventually reaped a variety of benefits from journal subscriptions and trade associations, they often joined in the first place because they desperately needed both technical and financial help with their business. Laundry ownership seems to have been a magnet for the inexperienced. Running a laundry gave every appearance of being relatively easy to learn, a means through which an enterprising young man (or very occasionally, a middle-aged woman) could acquire at least a comfortable living, if not a fortune. Like retail businesses, small laundries could be set up in neighborhoods or near city centers convenient to the owner's home. Specially designed buildings were not necessary unless one planned to buy the very largest machines. A steady supply of water and sewerage to remove wastes comprised the only critical prerequisites. Laundries often filled multistory warehouse or factory buildings originally designed for other purposes. Some small laundries were run out of the basement of the proprietor's home and depended upon the labor of family members. Though machinery made the processes more efficient, an initial investment in a full array of machines was not absolutely necessary nor, in some cases, advisable.

Fashion also played a role in attracting investors. From the 1880s onwards, laundries seem to have been one of those business opportunities–like the chicken ranch of the 1930s or the burger franchise of the 1980s–that caught the imagination of the public. Before the 1920s, when industry leaders began realizing that poorly run laundries reflected badly on the industry as a whole, this perception was often fed by trade literature that encouraged the unwitting to go into the laundry business, promoting the idea that laundries were good investments even for the inexperienced. As laundry owners became more forthcoming, their warnings opened a window onto the complications of the business of laundry.

In 1935, a British trade journal published an article entitled "Salvaging Small Laundries." The article could have served as warning to several generations of would-be laundrymen, beginning with the hapless Mr. Allan. The author's advice, which could be summarized as "Don't buy a laundry unless you know what you're doing," was presented as a cautionary tale. "An old school friend of mine in the south bought a small laundry business recently and, when it was too late for me to prescribe preventative medicine, he came to me for advice." The author described his visit to the premises, where he found poor equipment, a clammy cellar location, and account books that revealed a long history of financial losses. "How much better," he told his friend, "if you had bought a tuck shop, a tobacco shop, or a newspaper agency! Anything but a laundry business. Sweets and tobacco are commodities with an approved standard of excellence, whereas the best laundry is suspect!" He went on to explain why: "In a shop the issue would have been simple. Success would have depended on your ability to sell. Here, you will be required to have multiple genius. You must be engineer, chemist, psychologist, production expert, publicity manager, and general labourer."[46]

Factory inspectors noted that the rapid development of technology contributed to the problem. Lucy Deane, one of the British "lady inspectors," stated in her report for 1900, "I question whether in any other trade in the country so much ignorance of elementary matters connected with machinery is to be found." She went on to explain that "to have a gas engine that will drive the wringer and the heavy old 'box' mangle is the ambition of the smallest 'laundry.'" She found owners and workers dealing with newly acquired secondhand machinery had only the vaguest idea how to operate it. "The needful experience," she wrote, "is often brought at some risk."[47]

Getting and Using Machinery

The infrastructure of the industry—both trade associations and equipment makers and suppliers—was oriented towards dealing with (and occasionally taking advantage of) these smaller business owner-managers as well as the largest, most professionalized factory laundries. Machine makers like the Troy Laundry Machinery Company carried a variety of equipment for the largest to the smallest laundries.[48] Machine dealers like Armstrong also depended on the secondhand market to get rid of equipment that did not prove satisfactory for their largest clients.

The laundry owner could look for equipment through a number of sources. Trade journals featured the often elaborate ads of national manufacturers. The classified sections in the backs of journals offered used machinery, sometimes from failed laundries or laundries that were upgrading, put up for sale by companies that specialized in refurbishing secondhand machines. Through these ads one could also buy an entire laundry, complete with equipment, discreetly advertised with anonymous descriptions that made it impossible to identify the specific laundry without contacting the seller. Managers, engineers, forepersons, and the occasional ironer advertised for positions in these pages, suggesting that there was national (and perhaps international) traffic not only in equipment but also in people.[49]

Buyers could also consult trade catalogs. These varied enormously, from single-page brochures for a patented device to glossy volumes that could run up to fifty pages and featured elaborately detailed descriptions of goods offered for sale. During the early years of the industry, editors and contributors to the trade literature were often machinery manufacturers. A. M. Dolph, the first editor of the *National Laundry Journal,* also founded the American Laundry Machinery Company (despite his ownership of that business, he seemed willing to accept ads from other manufacturers).[50]

Machinery companies varied as widely in size and emphasis as their customers. The largest firms, like the Troy Laundry Machinery Company, had established sales offices in a half-dozen countries by 1910. The largest British companies, Summerscales and Sons and Manlove and Alliot, sold not only laundry machinery but also industrial cooking machinery, agricultural equipment, and other related devices. Directories for cities like Chicago also featured the names of many smaller companies that catered to a local market.

Machinery could be ordered directly from the manufacturer or through a dealer. Dealers could also supply replacement parts, particularly gear wheels and mangle rollers padded with special materials. There was a need for such parts, as belts broke, machine housing split, and factory inspectors could demand the installation of guards on machines or fencing around shafts and belts.

Owners and managers of large-scale operations and laundrymen that lived in urban areas could also inspect machinery at trade fairs and exhibitions. The editor of the *Laundry Journal Diary* touted the virtues of the second annual Laundry, Accessory, and Sanitary Exhibition in the Agricultural Hall at Islington. He noted with pride that the *British Trade Journal* had declared "the laundry exhibition was one of the most successful Trade Shows ever held at the Agricultural Hall."[51]

Once purchased, machinery had to be installed, operated, and maintained. It often required adjustments and sometimes took a certain "knack" to operate. Of course, no amount of tinkering could make a poorly designed machine run well, and an ignorant laundry owner might not be able to ascertain whether the problem was inherent in the machine. The same was true of laundry chemicals–soaps, bleaches, starches, and blues. Trade journals were of little help in this instance. They depended too much on advertising revenues from machine and supply dealers to honestly critique individual machines as they entered the market.

Advising purchasers of new machinery on its operation and providing what might now be called "technical support" was much more common in the British context than the American. British owners who desired the services of an expert could turn either to machinery manufacturers or to firms that specialized in laundry engineering or laundry architecture. A 1908 trade directory carried the names of many such providers. James Armstrong and Company, one of the largest such firms, advertised themselves as laundry engineers (they also acted as "sole European agents" for the American Troy Laundry Machinery Company). One could also consult T. Herbert Pearson for architectural advice or employ one of the smaller engineering firms that might not have Armstrong's potential conflict of interest.[52]

The "laundry engineer" was a key figure in the British laundry trade.[53] The engineer provided technical advice, floor plans, and even turnkey facilities for fledgling entrepreneurs, often acting on behalf of the machinery manufacturer. Large laundry machinery manufacturers like W. Summerscales and Sons, would even provide purchasers with advice about bookkeeping and the proper packing and wrapping of finished goods.[54]

In a brief article entitled "Laundry Engineers," the *Laundry Journal Diary* (1894) stated: "We have often expressed a conviction that the only hope for the so-called 'cottage laundress' was in an attempt to improve their methods so as to conform to the greater demands made by sanitary advances of the day. That many of the 'smaller fry' have seen this, is abundantly evidenced by increased patronage accorded to laundry engineers." The 1908 *Laundry Trade Directory and Launderers' Handbook* went farther, advising the entrepreneur, "Don't convert a building or a hand laundry into a power laundry without expert opinion; get only the machinery you require, and be sure the working of the laundry does not create a nuisance."[55]

American machine dealers, however, often assumed that equipment would be set up by an owner-manager. The text describing a very large mangle in a 1901 Troy Laundry Machinery catalog states that "the directions for setting up and starting the mangle are very complete, and we feel sure that any one having some mechanical ability would have no difficulty in putting it in perfect running condition."[56] In the absence of formal technical support, American entrepreneurs turned to trade catalogs and advice from managers experienced in the business. Mrs. Lillian G. Barrow wrote to the editor of the *National Laundry Journal* in October 1902 about her travails in starting a small laundry. Her story was in many ways typical, but it was probably printed in part because of the novelty that she was a woman from the South entering into laundry ownership. The three managers she had hired each had stayed "just long enough to get a good trade started and under different pretenses left me with clothes in suds and starch and my pocketbook so light I had to put a flatiron on it to keep it from floating off." Eventually she had learned to run the entire plant herself: "My son has gotten to be a pretty fair ironer, though he is only 14 years old, and I have a good

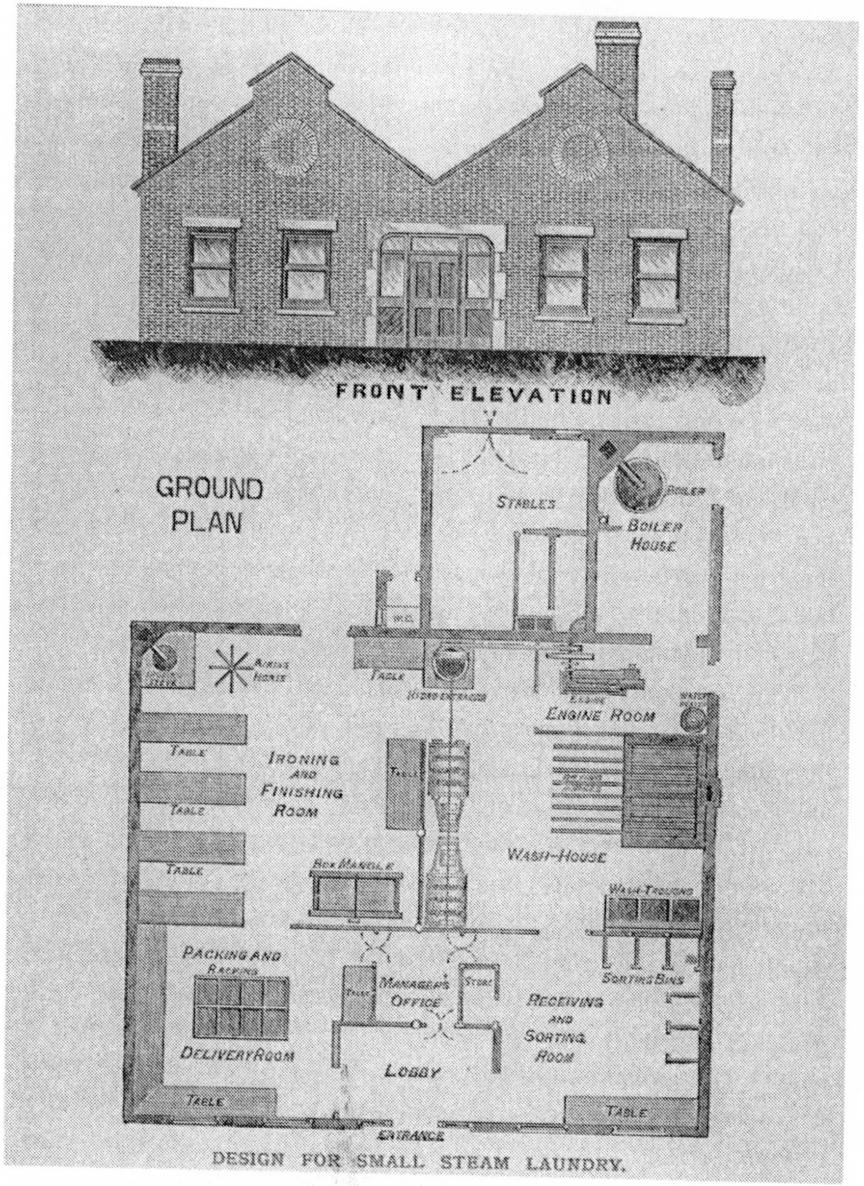

Fig. 5. Large British laundry machinery manufacturers could even provide would-be laundry owners with turnkey laundries. W. Summerscales and Sons offered this example in their 1897 catalog. W. Summerscales and Sons, *Laundry Machinery* (London, 1897), 17. Courtesy of Science & Society Picture Library, London.

force of colored girls who do the other work. I do a little of all of it, and have entirely overcome the fear I had once of not giving satisfaction."[57]

While the importance of machinery might have been obvious, would-be laundry owners often overlooked the importance of choosing an appropriate building. The former skating rink below a stable that had been selected by the directors of the American Steam Laundry Company was an only slightly exotic variant on the wide array of buildings used to house laundries. Proper drainage and ventilation, a good source of light, floors sound enough to hold machinery, and fireproofing were all among the important requisites of an adequate building. Factory inspectors and workers were often the first to notice the risks inherent in some locations. One anonymous worker wrote in response to a New York Women's Trade Union inquiry that she feared for her life working in a loft making paper boxes above a laundry. She said the greatest risk was the boiler, for "the engineer says [it] is in a very bad condition, likely to burst at any time." She also worried that "the heat of the mangles is bound to set fire to this factory before long."[58]

While machinery was purchased only occasionally, laundry operators dealt with supply houses on a day-to-day basis. "Drummers" regularly made the rounds, bringing soap and chemicals, packing material, and padding for the ironing machines. Paul Siu researched his University of Chicago dissertation on Chinese laundries by working as an agent for a laundry supply house. From this vantage point, he became intimately acquainted with the workings of the laundries.[59] Some supply houses were very large. The John T. Stanley Company of New York, which wholesaled laundry supplies, employed a sales force of 150 people.[60] Like many small businesses, laundries occupied a middle position on the continuum between production and consumption. While customers saw them only as producers of clean laundry, laundry owners were also customers for a host of subsidiary industries—from new machinery to soap flakes and shirt boxes. Once in place, this network of supply generally ran smoothly, but its decentralization and small scale could also leave smaller laundries vulnerable to dishonest machine dealers and erratic suppliers.

The Competition

Laundrymen's keen awareness of outside competition helped fuel not only an attention to the inner workings of their laundries but also a sense of solidarity within the industry. American anxieties fixated on the so-called Chinese laundries that had sprung up in the 1880s as Asian immigrants were pushed out of other employment by restrictive legislation. In a number of American cities, the competition from these small laundries was real. For instance, in 1880 nearly 10 percent of San Francisco's Chinese population was engaged in laundry work.[61] By the early years of the twentieth century, Chinese-American entrepreneurs had spread eastward, establishing themselves in Chicago and in some East Coast cities. By employing relatives who worked for room and board, owners of these laundries provided inexpensive services competitive with power laundries. Some consumers also found them appealing because they purported to carry out all processes by hand, avoiding the risks that machines posed to clothing. By the time of the 1909 census, these Chinese laundries had also diversified. Some invested in machinery, others subcontracted the heavy work of washing to larger laundries and then hand finished themselves. Some even hired white, female employees, prompting a few local governments to pass legislation, supposedly to "protect" these women by barring them from this kind of employment.[62]

Industry defenses against this kind of competition sometimes made for strange alliances. In 1908, the powerful San Francisco local branch of the laundry workers' union allied with the laundry drivers' union, the proprietors of steam laundries in that city, and French laundry owners to form the "Anti-Jap Laundry League." Racial allegiances overcame class sympathies as these disparate and often antagonistic groups pooled their resources to set up an office to drive Japanese laundry owners in the Bay area out of business. Their method was to follow the laundry wagons of the Japanese laundries, note the customers' addresses, and then send threatening letters to the patrons. Should this tactic fail, a woman employed by the League would be sent out to have a "little chat" with the housewife and elicit from her a promise that she would not betray her race by continuing to patronize the competition.[63]

Women who took in washing or appeared on an employer's doorstep to wash in the kitchen on Monday morning also continued to be a social reality well into the depression. The waves of African American migration from the South into Chicago brought with them experienced washerwomen who found work in the homes of the well-off. It is very difficult to ascertain how many women actually did this kind of work. In 1900, the census counted 328,935 laundresses. If only 82,000 worked in power laundries, an enormous remainder worked either in hand laundries or in homes. Numbers are also skewed by the fact that enumerators made varying decisions about how to count employment.[64] A 1911 study by the Immigration Commission found that taking in laundry was the most common form of homework in the cities polled. These laundresses were classified as mostly "Negro," "Bohemian," or "Moravian."[65]

British laundry owners also worried about competition, though they identified different sources for it. They found the American hysteria about the Chinese amusing. Chinese laundries were almost unknown in London at the turn of the century. A few could be found in port cities (one 1908 publication counted thirteen in Birkenhead and sixty-nine in Liverpool),[66] but these were looked on as a curiosity. Instead, the British feared competition from washerwomen. Trade journal rhetoric about these women echoed the anti-Chinese bombast of the American laundry journals. Both suggested that their rival workers were unclean and untrustworthy. From time to time, the British journals employed experts to decry these institutions. For instance, in the March 15, 1899, edition of the *Laundry Journal,* a Dr. Harris described visiting a laundress who was packing laundry in the same kitchen where a child lay sick with scarlet fever.[67] Laundrymen also waged a legal battle to keep laundresses from using public facilities and from persisting with traditional methods in an urban environment. One longstanding fight involved the laundrymen working to prevent laundresses from drying their laundry on the Common at Tunbridge Wells.[68]

The most perplexing problem for both British and American laundry owners, however, was the housewife. Though women frequently testified that they loathed laundry work more than any other domestic task, surprisingly few seemed willing to hand over the family wash. Some did the work themselves to save a few pennies, others employed a washerwomen or a Chinese laundry. Those that did patronize laundries often used them sporadically or only for particularly difficult-to-wash items like sheets and men's shirts. Since these competitors were also potential customers, neither vilification nor price competition was an option. Laundry owners might have looked to other industries that had industrialized domestic tasks, but they would have found few analogies and little inspiration. This peculiar competitive relationship with the customer was their own special cross to bear.

⁂

Those that entered into the laundry business entered into a complex world. They faced the multiple challenges of choosing and using technology, managing workers, and dealing with consumers. The business of laundry encompassed far more people than just those who labored within the laundry. It created a kind of community bound together by common experience and common expectations. This community crossed national borders, actively exchanging ideas and assessing other members. As the community tried to meet these challenges, differences of scale and status sometimes became more important than those of nationality.

3

Inside the Laundry

DURING THE LATE NINETEENTH CENTURY, LAUNDRIES WERE NOT only organized into an industry but also underwent a long period of technological change. By the turn of the century, observers were describing many of them as factories. These laundries were characterized by mechanization, division of labor, and increase in scale, and they were distanced both materially and symbolically from the kitchens and workshops where washing had long been carried out.[1] Social investigator Elizabeth Butler, for instance, compared this transformation to the evolution of the textile industry that had begun nearly a century before. "The solitary washtub and red-armed washwoman as industrial types are passing as surely as individual loom and shuttle have passed," she told her readers.[2]

The gradual introduction of new machinery served as the most concrete indicator of change. Between 1880 and 1900, large washing machines and mangles heated by steam became widespread, mechanizing the least skilled and most physically demanding parts of the process. Between 1900 and 1910, a number of innovative ironing and starching machines were introduced, mechanizing what had previously been considered the most skilled jobs in laundries. Butler found widespread evidence of laundries adopting new devices and new methods. Inside the low stone buildings housing many of Pittsburgh's power laundries, she encountered a wide variety of machinery: washers, mangles, and "steam ironers of a dozen kinds." Even the division of labor, Butler noted, "is carried as far as in a factory. There is specialization for speed."[3]

Lucy Deane, one of the British government's "lady inspectors," noted similar changes in her report to Parliament. She claimed the industry had undergone an extraordinary metamorphosis since the inspectorate had begun its visits in 1892, eight years previous. Even the smallest proprietors now eagerly mechanized: "At one time, it was only in a few large steam laundries that machinery was to be met with. Now it is no uncommon thing to find a row of houses in separate occupation, the backyards of each roofed in and packed with laundry machinery, all driven by an engine installed at the end of the row."[4]

Deane's observations also suggest mechanization was widespread but not evenly distributed. Almost all laundries carried out the same sequence of processes: Drivers or receiving desks collected bundles, passing them to a checker who marked the contents off against a list. Sorted by type and fabric, goods moved through a series of departments, or, in very small laundries, individuals. There, specialized workers washed, dried, starched, and ironed the items–collars and cuffs, lace curtains and dinner napkins–before re-sorting and packing them to be returned to their owners. Machinery and division of labor could be used in all or none of the steps in the process. Most often, small-scale owners bought secondhand machinery or did without, relying on traditional tools and the deft hands and strong backs of their employees. Meanwhile, large, well-capitalized laundries bought new gadgets as they came on the market, banking on systematization and economies of scale to provide a profit.[5]

Wide variations in the degree of mechanization within regions make it difficult to compare the rates at which the British and Americans adopted new technology. It is clear that the flow of machinery was from the United States to Britain, but the largest British and American laundries bore more resemblance to one another than to their smallest competitors. The records of Bird's laundry suggest British firms were importing American machinery as early as the 1880s. By the turn of the century, markets were similar enough that the largest companies could sell their products abroad with a minimum of alteration.[6]

This chapter describes the work processes one could have expected to find in a medium-sized or large power laundry between approximately 1895 and 1910. I follow a convention adopted by reformers like

Deane and Butler, who believed that the problems of laundry employment were technical as well as social. To understand those problems, readers have to understand how machinery, chemicals, buildings, and techniques did and did not work.

The following tour also helps make clear why mechanization, division of labor, and economies of scale were mixed blessings for laundry owners. Despite a growing resemblance to manufacturing, this was industrialization with a difference. Economies of scale and mechanization, when applied to laundry work, brought with them constraints unfamiliar to manufacturers of bicycles or penny nails. All these strategies had to be utilized within certain fundamental limits, so as not to jeopardize the laundry industry's intimate association with many of its customers. For instance, the greater a laundry's volume of business, the harder it became to keep track of individual items and to reassemble them into a single bundle to be returned to their owner in a timely fashion. Finding a missing collar among one hundred was easier than finding one among the twenty thousand that passed through some laundries in a given week.[7] Laundries also differed from manufactories because laundry owners could not control the quality of material entering the laundry. Mechanization could increase the risk of damage to delicate or shoddy fabrics. A worker doing handwork might more easily adjust her techniques to preserve a threadbare tablecloth than an operative could adjust a machine.

Done well, mechanization could remain invisible to consumers. In the laundry industry, "that looks like it was done on a machine" was not a compliment. Although laundry patrons found mechanization appealing in the sense that it took their clothing out of the potentially unclean hands of washerwomen, they also worried that machinery might destroy their clothing. Laundry owners disagreed about the reality of this perception. Some pointed out that many so-called hand laundries sent clothes out to be washed by machinery while the customer remained in blissful ignorance. Others agreed that new chemicals and machines were potentially harder on clothes than a washboard and soft soap, but they also pointed out that rising standards had encouraged the use of these harsher techniques. Consumers expected their clothes to emerge from the laundry absolutely spotless no matter how stained and filthy they had been when they were dropped off.[8]

On the other hand, mechanization offered a solution to the problems of rising wages, attracting and keeping skilled workers, and dealing with unexpected amounts of work. As legislatures on both sides of the Atlantic imposed laws against subjecting women workers to overtime or nightwork, laundry owners could no longer keep workers all night on a Friday to finish unexpected work. Machinery also offered a way to have excess capacity without hiring excess workers.

The material in this chapter is drawn from a variety of sources, including trade literature, trade catalogs, reformers' accounts, and British working-class autobiographies. It encompasses several very different perspectives on technology and technological change. Employers and machine manufacturers described technology primarily in terms of its usefulness in solving problems of production. Likewise, they portrayed social relationships–among workers and between workers and employers–in terms of the way those relationships helped or impeded the production process.

While reformers and factory inspectors appreciated (and occasionally waxed rhapsodic over) advances in production, they took an interest in laundry technology primarily as part of the environment in which working women and men spent most of their waking hours. Educated in an era of investigative sociology and muckraking, they wanted to draw their middle- and upper-class readers into the world of working-class women and provide an understanding of the squalor, heat, and danger of their work lives. Before the 1920s, some also described their own experiences of having worked undercover in laundries. Inspectors, meanwhile, also developed considerable expertise in laundry machinery (better than many employers, in their estimation).[9]

The experiences of workers offer a third viewpoint. With the exception of some British working-class autobiographies, most workers' accounts have been filtered through reformers' reports.[10] The middle-class women who recounted their experiences masquerading as workers or reconstructed interviews with workers of necessity shaped their accounts to further their cause of reform. Still, even though much of it is secondhand, this perspective is important. For workers, the experience of the technology was immediate. They expected to adapt themselves to the workplace environment and the machinery. Technological innovations could either ease the strain of a particularly dreaded job, or they could worsen it by facilitating expectations of speed or by straining one particular part of the body. Everyone in the laundry suffered some of the same maladies. The combination of heat, damp, and constant standing led to swollen feet and ulcerated legs. Burns were also common.[11]

The Work Environment

The workday in most laundries began early: between six and eight o'clock in the morning. First in the door were the washmen and the person who stoked the engine and fired up the boiler, usually a specialized worker often referred to as the "engineer," but in smaller businesses it was just as often the owner or manager. In laundries with gas-heated irons or mangles, the gas had to be turned on and lit.[12] Elizabeth Butler described the early morning scene in the laundry districts of Pittsburgh: "If you walk along Chartiers Street near seven in the morning, you may see the girls coming down the street to their work . . . the whistle from a neighboring factory shrills out the hour, there is the slow sound of an engine starting, a gathering whir of belts and wheels, and the last girls disappear to take their places at the machines."[13]

Workers arriving in the morning hung their coats or shawls on pegs in the corner of the workroom and sometimes exchanged their shoes for a larger pair (or, in at least one laundry, bedroom slippers), which would accommodate the swelling that came with heat and constant standing.[14] Some larger laundries had dressing rooms, and a few employers even provided lunchrooms (or messrooms, as they were called in Britain). In most places, workers thought themselves fortunate to have a clean, functioning toilet and a source of potable drinking water. Factory inspectors fought battles to secure and preserve such resources.[15]

For the uninitiated, the fetid, humid atmosphere of most laundries was a shock to the senses. The washing machines started first in the morning, filling the basement and floors above it with steam and the smell of chemicals. For some visitors, this paled in comparison with the odor emanating from the sorting room. One physician described it as so thick and foul that "it seems to cling to one's lips till one tastes it."[16]

The experienced and those bent on self-preservation, unbuttoned the collars of their shirtwaists and pinned their hair out of the way of snatching belts and rollers. Almost everyone had heard stories of young, vain women who were caught and scalped (or worse) because of their long, loose hair. In the extreme heat of the laundry, vanity was sacrificed to comfort. Some might come on their first days in their best finery, but that was soon abandoned for comfortable clothing. "A woman in a laundry," stated one laundryman, "seldom wears that thought-to-be-necessary-garment: the corset."[17]

Unlike many factories, however, laundries were not unbearably loud. Hand laundries had the hush of workshops, with only the swish of the dolly in the washtub and the hum of the ironers leaning into their work. The addition of machinery added the rhythmic grind of gearing, the whispering of belting, and the continual hiss of escaping steam—but this did not compare to the unbearable racket of textile factories and other manufactories. Noise was mostly an annoyance. Being able to talk was one of the things that many women enjoyed about working in laundries as opposed to other kinds of factories. Despite work rules and managers who forbade talking and singing, many workers did so anyway. One wag suggested that "many lung complications have been traced to the inhalation of soapy steam, but if laundry employees talked less over their work, the danger of this would be considerably lessened."[18]

The Work Process

Soiled goods arrived at the laundry in one of two ways: they were either delivered to the front counter by customers or brought in by drivers after their rounds to houses, apartments, hotels, and barbershops. Laundry owners and managers paid special attention to these interactions with the public. One business handbook advised laundrymen that the "office is the figurehead of the business. The affluent laundryman might embellish the office with "stained glass, paneled ceilings, carpeted floor . . . and silverplated show cases," but it was far more important to have the office "at all times neat, sweet and clean, so that the most fastidious person who entered would never see or *smell* anything offensive" (emphasis in the original). In particular, the office should be located as far from the sorting room as possible.[19] Most laundries also put limits on what they would accept. Most importantly, they rejected "suspicious looking bundles" that they feared might spread disease. A model laundry ticket from 1920 also suggested to proprietors that they refuse responsibility for the various containers, including suitcases, that were left with the dirty laundry.[20]

Pickup and delivery services had the advantage of keeping customers out of the laundry and the disadvantage of vesting the driver with a great deal of power. Because drivers handled money and negotiated complaints, they were sometimes seen as a surrogate for the laundry owner or manager. A surly or dishonest driver could ruin his em-

Fig. 6. A well-turned-out wagon or truck was a point of pride and an important advertisement for laundries. The high-wheeled design of this wagon helped the driver, John Youngman, negotiate the rutted, muddy roads around Newcastle. From Bird's Laundry, Tyne and Wear Archives, Newcastle-upon-Tyne, © 1920. Courtesy of the Tyne and Wear Archives.

Fig. 7. In narrow, urban streets, British laundry owners sometimes relied on motorcycles or bicycles for deliveries. The caption for these bicycle illustrations read "Two types (one with pneumatic tyres for express delivery) of Warwick's 'Monocle' tricycle carriers specially built for laundry work." G. Cadogan Rothery and H. O. Edmonds, *The Modern Laundry*, vol. 2 (London, 1902), 225. Courtesy of Hagley Museum and Library.

ployer's reputation. If the driver was charming and energetic, however, he could bring in new customers. The replacement of horse-drawn wagons with motorized trucks accentuated this role. Furthermore, owners or managers no longer had to hire someone who was skilled at handling both people and horses.[21]

For more prosperous businesses, a handsome laundry wagon (later truck) was the laundry's most conspicuous advertisement in the community. Company photographs of large laundries often show a fleet of wagons or trucks lined up in front of the building or, perhaps more modestly, a single well-turned-out wagon. A few British laundries also used bicycles or scooters to hurry through the same narrow alleys where washerwomen sometimes wheeled handcarts or wheelbarrows.

"Laundries are judged by the teams they have on the street," advised one laundryman. "Never send a team on the street you would be ashamed to ride in yourself." Realistically, as in other businesses that used delivery wagons, there were some laundry owners who, out of ignorance, because of a shortage of funds, or simply because they did not care, nevertheless probably sent out lame or underfed horses. Another advantage of a truck was that it did not need to be fed and cared for–a worrisome expense when business was slow.[22]

Keeping Track of the Laundry

Whether laundry arrived by delivery wagon or directly from the customer, it went to the checker first. Checkers played a crucial role in the complex process of keeping track of individual items in a customer's bundle so that they could be reassembled at the end of the process. The checker filled out a laundry list or compared the contents of bundles against a list provided by the customer. A favorite scam of dishonest customers was to name more items than were actually in their bundle. The customer would later demand the cost of replacement for these phantom items.[23]

Checkers had to have good bookkeeping skills, legible handwriting, and a presentable appearance in the laundries where they met the public. In very small businesses the owner might handle this duty, because a great deal of customer goodwill could be lost on improperly managed transactions and money squandered paying claims on lost or

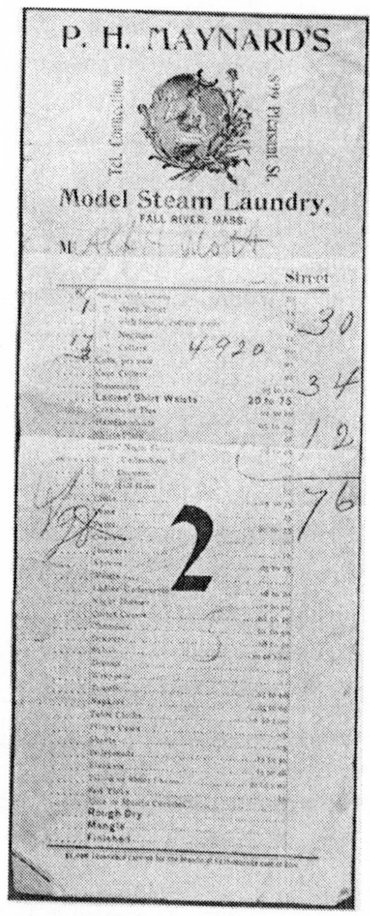

P. H. MAYNARD'S

Model Steam Laundry,
FALL RIVER, MASS.

Street

Ladies' Shirt Waists 20 to 75

Rough Dry
Mangle
Finished

Fig. 8. Laundry lists were filled out either by the customer or by the checker (author's collection, ca. 1900).

damaged articles. After the turn of the century, most checkers were female. The women Elizabeth Butler interviewed in Pittsburgh had only recently entered what had been an all-male job. As laundries grew in scale and other processes became mechanized, checkers and markers set themselves apart from other laundry workers, resembling clerical workers more than manual laborers.[24]

Marking came next. Markers used ink or thread or one of a number of patented marking systems to place an identifying code in an inconspicuous place on the clothing. Small laundries often identified goods by writing the customer's initials or last name in indelible ink. Larger laundries developed complex schemes to deal with the thousands of items passing through their doors every day. This process was later aided by key punch machines, which eliminated the problem of illegible handwriting. Companies like the National Marking Company of Cincinnati, Ohio, turned out sophisticated systems for large laundries. Most laundry owners considered marking a necessary evil. No marking system was without its drawbacks. Ink left an unsightly border around frequently laundered items such as handkerchiefs and napkins. Pins tore loose in washing and ironing machines. Colored thread was time-consuming to apply and remove. Because there was no standardized industry-wide system for marking, customers who traveled or changed laundries frequently could end up with clothes embellished with a wide variety of laundry marks.[25]

Prior to washing, most laundries separated their work into three general categories: white, colored, and black. Flatwork–sheets, towels, and napkins–were most easily managed, while men's shirts and collars (the bulk of the business for some laundries) constituted another category. Sorters placed small or delicate items in net bags so as not to be lost. Colored goods, flannels, woolen underwear, and silks that could be damaged by harsh processing were also separated.[26]

Mechanization coincided with laundry owners' efforts to expand their market beyond shirts and sheets into what was often called the "family wash." Although mechanized laundry processes were best suited to white cotton goods, particularly flatwork, many neighborhood laundries would accept almost anything. British laundries, for instance, did a sizeable trade in wool flannel trousers. Others specialized in the "carriage trade," taking in the gentry's finery, including difficult-to-wash silks and laces. One launderer's breakdown of the 12,279 lbs. of work handled in his plant included a large amount of colored children's clothing, curtains, coats, and sixty-seven blankets.[27]

Sorters had to be knowledgeable about types of fabrics. Adulteration was a constant concern. The practice of weighting silks with metal oxides, for instance, had disastrous consequences in the washroom. Some black silks owed as much as twenty-five percent of their weight to adulterants. Manufacturers gave distinctive textures to other fabrics with coatings that disappeared or turned yellow when washed. Laundry owners and managers also worried about the substitution of one fabric for another, mercerized cotton for more expensive linen, for example, and the mixing of fabrics such as cotton and linen.[28] Through various publicity campaigns, they hoped to shift public blame for damage to adulterated fabrics away from the laundry. The British trade association published a booklet entitled *The Lady, the Linen, and the Laundry,* which dealt with "the difficulties of the modern laundry proprietor through the cheapness of manufactures."[29] The American association, the LNA, waged a long and bitter fight for fabric content labeling in clothes.[30]

A few laundries also did dry cleaning. Though the principles of using benzene, gasoline, or other solvents on nonwashable items were widely known from the 1890s, these chemicals mostly augmented a variety of substances already used for spot-cleaning. Re-dying remained a more common strategy for "reviving" (*cleaning* would be a misleading word in this context) garments such as men's suits. Early dry cleaning was extremely labor-intensive, and the chemicals involved were dangerously volatile. Until the introduction of effective machines and stills for recycling the used fluid, this sort of work usually existed only as a small sideline for a few laundries that were willing to spend time fussing and rubbing.[31]

Washing

Once sorted, goods were placed in trucks or large baskets and taken to (or slid down a chute into) the wash house or "kitchen" of the laundry. Despite all advice to the contrary, laundry owners habitually located washrooms in the basements of retrofitted buildings. They saw rising steam and windowless workspaces as a lesser evil than the possibility of rotted floorboards, collapsing ceilings, and flooding caused by washing on upper floors.[32] Inside the wash house, men or boys loaded the laundry into hatches on the tops of large cylindrical washers lined up against the walls. Everyone's laundry went in together–something that disturbed some sanitary advocates and unsettled people with the thought of their private things mingling with those belonging to strangers. In a fit of Victorian propriety, one 1889 manual suggested that "Ladies' linen, as a rule, should be treated apart in the Private Washhouse"–a prescription that found few takers.[33]

The washing machines most familiar to turn-of-the-century laundrymen consisted of an external wooden cylinder and an internal perforated metal cage. Before the introduction of galvanized metal, brass was the preferred material because it did not rust. By 1910, machinery manufacturers were offering all-metal machines in increasingly larger sizes. Some hand techniques lingered in smaller laundries. In 1896, Charles Booth found "dollymen" in some London laundries, continuing to use a wooden dolly stick to agitate clothing in barrels and tubs. To cater to this clientele, British manufacturers continued to offer traditional tools along with new machines.[34]

Fig. 9. In the 1890s, British trade catalogs still offered traditional tools for hand laundries alongside the most up-to-date machinery. From W. Summerscales and Sons, Ltd., *Laundry Machinery* (London, 1897), 96. Courtesy of the Science & Society Picture Library.

Once the goods were loaded into the washing machine, the washman opened the valve, filled the drum about one-quarter full, and added a bucket of soap. Until the introduction of detergents in the 1930s, all soap was manufactured by creating a chemical reaction between fats and an alkali. Workers often made soap on the premises. In 1893, one large New York laundry gave its recipe to *Scientific American* as follows: 500 lbs. of tallow, 10 lbs. of caustic potash, and 70 lbs. of caustic soda. The washmen diluted 75 lbs. of this mixture with 600 to 700 gallons of water and heated to between 100 and 125 degrees Fahrenheit. The resulting mixture was scooped into the washing machines with a bucket. Other laundries bought bar soaps or soap chips in barrels and dissolved them in water, usually in huge tanks in the corner of the washroom.[35]

Most machines were belt-driven, running off a power shaft suspended from the ceiling, until the use of electricity began to spread during the 1920s. As the washer filled, the washman threw a belt onto gearing at the end of the cylinder and a reversing gear caused the cylinder to rock. Machines agitated the clothes with suds from ten to forty minutes, though if inadequate soap or very hard water was used the process might be prolonged for more than an hour. Washing involved not only the initial sudsing but also multiple rinses, to which the washmen added various chemicals–most commonly a bleaching agent such as carbonate of soda or chloride of lime, neutralized by a "sour" such as oxalic or acetic acid. Bluing, or "blue," added to the final rinse combated the yellowing effects of the other chemicals.[36]

A single individual often tended a number of machines, keeping track of the amount of time goods had been in machines and opening up the hatches during the process to judge sudsing, to drain off washing "liquor," as it was sometimes called, and to add rinse water, bleach, or bluing. In laundries without strict rules about procedures or a watchful manager or foreman, the washman exercised considerable judgment in timing washes and measuring amounts of chemicals to be added. Miscalculations or lapses in concentration could result in damage to goods–damage that did not always show up until the garment was dried.[37]

After a sufficient period of time, the washman disengaged the belt, opened a valve, and drained the water off a final time. He then pulled the wet goods out of the machine and loaded them into a basket or truck, which was arduous, back-breaking labor. These cylinder washers had no "spin cycle," so whatever water did not drain out of the machine by gravity remained in the wash, only to flow out onto the floor, soak the clothes of the washman, and fill the air with steam and chemical odors.[38]

Washing was wet work. Under the action of water and chemicals, machines with wooden drums eventually rotted, sieving water onto the floor. These drums also splashed water when opened. Many basements drained so poorly that water and chemicals stood in pools around the washmen's feet. In Britain (where adequate drainage was required by law), one inspector noted that "it is not at all uncommon therefore to find that the yellow and foul water from a row of tanks or washing machines at one side of a wash-house flows all across the floor and over the feet of the workers before eventually reaching the drain." Washmen endured constant heat and swollen or ulcerated legs from the persistent damp. They coped by stripping to their undershirts, putting on clogs or waterproof boots, and taking an occasional nip to ward off the discomfort.[39]

Although the washing machines were the exclusive province of male workers, some laundries also employed a few women to wash delicate goods by hand. Elizabeth Butler described these handwashers as "women whose strength had gone at other trades." She continued, "For $1.00 a day they work ten hours over the tubs, at flannels and socks and bits of finery that would need to be handled by an extra machine if they were machine done." One of the managers told Butler that "you can't get a young woman to do this work."[40]

Drying and Finishing

Excess water was removed from the still-dripping laundry using a device called a *centrifugal extractor*. The extractor was a drum-shaped device with a loose perforated interior basket. The operator packed washing in as evenly as possible, covering the basket with a towel or lid. The belt was then thrown on and the extractor began to spin, removing water by centrifugal force. Extractors came in two different versions. They could be driven either from underneath, using a countershaft which turned the basket at three to four hundred revolutions per minute, or from overhead, using a friction gear which provided more rapid acceleration (thirteen to fourteen hundred revolutions per minute) but which was also far more dangerous.[41]

The wise laundry owner tightly anchored the extractor to the floor to prevent it from spinning off into a wall or unsuspecting bystander. Even when securely anchored, however, extractors presented other dangers. After the belt was removed, the inner drum continued to spin for a few moments. Before safety legislation mandated extractor lids that locked, one of the most common fatal accidents in a laundry resulted from a worker's arm or length of hair getting caught in the rotating drum.[42]

Because it had an ambiguous function, somewhere between washing and drying or ironing, the extractor was the only relatively gender-neutral piece of technology in most laundries. In Britain, until child labor laws were strengthened, laundry owners often described the operator as "that genial youth, the extractor boy."[43] Women and washmen also operated extractors.

Goods emerged from the extractor twisted and knotted. A few laundries employed a wooden tumbler to shake them out.[44] More often laundries employed young girls at low wages for this task. Because shaking was the easiest laundry job to get, it was the one most often taken by reformers seeking undercover experience. Journalist Rheta Childe Dorr vividly described her brief stint shaking: "Often the ropes were so tightly wound that the untwisting process bruised our hands and broke our fingernails. Worst of all was the speed, on which the forewoman insisted, and the deadly monotony of the task. Bend, untwist, shake; bend, untwist, shake; bend, shake; bend, shake, hours unending."[45]

After shaking out, flat goods such as tablecloths and napkins went to be mangled (or calendered, to use the British term). The mangle pressed out wrinkles and dried the goods at the same time. *Mangle* is a confusing term, applying to several quite different devices. The most venerable of these was the box mangle, which originated sometime before the nineteenth century. The user wrapped flat goods around a roller or laid them in the bed of the mangle. She then moved a weighted box back and forth over the goods to press out the wrinkles.[46]

Fig. 10. Box mangles used pressure rather than heat to remove wrinkles. This power-driven model represents an unusual hybrid of old and new technologies. From G. Cadogan Rothery and H. O. Edmonds, *The Modern Laundry* (London, 1902), 143. Courtesy of Hagley Museum and Library.

Though box mangles were occasionally used in power laundries, by the 1890s they had been largely replaced by the steam mangle, or decouden calendar. Steam mangles came in two types, the chest mangle or the cylinder mangle, which were differentiated on the basis of whether heat was produced by heating the rollers themselves or from steam chests under the bed of the machine. Operators fed flatwork underneath a series of rollers padded with fabric (usually wool felt or a combination of wool and canvas) and heated either by steam or by gas flames. In British laundries, the term *calender* referred to several related devices. Before 1900, British laundries often used a decouden calender, similar to the mangle except that the metal drum was not padded and instead made direct contact with the fabric, imparting a slick, glossy finish. *Calender* was also sometimes just another name for a steam mangle. In British laundry catalogs, the American mangle is also simply called an *ironing machine*, probably to distinguish it from the older box mangle.

Because operators had to insert the flatwork onto the moving bed ("apron") as close to the first roller as possible to avoid skewing the goods entering the machine, burned or crushed fingers were a common laundry accident. (Machinery companies eventually started calling these devices *ironing machines* instead of *mangles*, partly to avoid the implications of the latter term.)[47] Workers told one investigator

The Trojan Mangle

Fig. 11. Steam mangles were used to iron flatwork such as sheets, tablecloths, and towels. From the Troy Laundry Machinery Company, *Illustrated Catalogue of Laundry Equipment* (Chicago, 1907), 109. Courtesy of Hagley Museum and Library.

they made decisions about how close they dared come to the mangle according to the quality of the work demanded. They inserted "boarding house work" with less care than "millionaire's work," which was shoved to within a quarter inch of the cylinder in order to guarantee it would be perfectly ironed.[48]

Like other devices, mangles increased in scale during the late nineteenth and early twentieth centuries. The Paragon mangle, made by the Troy Laundry Machinery Company, was fourteen feet wide, sixteen feet long, and weighed thirty-two thousand pounds. It came in sections so that it could be expanded or shortened according to the laundry owner's needs.[49] The largest mangles often had ominous names. The British Henrici Laundry Machinery Company produced a device called *The Annihilator.* The ad copy let the mangle speak for itself: "I am the annihilator. My name does not mean that I 'annihilate' the clothes, [or, presumably, the machine operators] but . . . it does mean that I can do more work . . . than any other three machines for the same purpose."[50]

Shirts, collars, underwear, dresses, and other cotton goods followed a different path, instead going to the starching room, drying closet, and ironing room. Starching was a time-consuming process requiring patience and judgment. Before the 1920s, most laundries made their own starch out of ground grain, usually wheat or corn, and water. The raw material came as a dry powder, which was put into a tub, covered with cold water, and worked with the hands until entirely dissolved. At the turn of the century, most English laundries used the starch in this form, while Americans boiled it first to make a gelatinous paste. In a well-equipped U.S. laundry, this mixture was boiled in a sealed, free-standing vessel called a *starch cooker* until it was judged to be of the appropriate consistency. Laundry owners unable to afford such specialized equipment simply cooked it in a pan or kettle on top of a stove.[51]

Starchers thinned the finished product with water to achieve the desired consistency and applied it, then they wiped the clothing by hand to remove any excess. The women in this job had to make judgments about the temperature, consistency, and amount of starch to be applied in order to produce an acceptable product. These choices, as well as which type of starch was used, also contributed to the color and type of finish imparted. Consumer demands about starching varied. They could specify not only "light starch" or "medium starch, please" but also the type of finish.

All kinds of things could go wrong during the starching process: the starch could be too thick or thin, it could "sour" in hot weather, and goods could emerge spotted or discolored. Like other chemical parts of the laundry process, starching was made more difficult by the fact that manufacturers had not yet begun to scientifically control products for consistency. By 1910, many devices had been invented to mechanize starching. Collars and cuffs passed through the hot starch on a belt. Shirt bosoms were starched on a bosom-starching machine. This saved the arduous labor of rubbing in the starch by hand, though most goods still had to be hand wiped. Goods that had been "hot starched" had to be dried once more so that the starch would not stick to the iron.[52]

Starched items and those destined for hand ironing were dried in a drying closet rather than under the heated rolls of a mangle. Into the 1920s, these closets were most often enclosed boxes with a heating element (pipes from the boiler or gas heat) and a blower. The simplest version obliged workers to walk in and out to hang up and remove clothes—a situation that horrified inspectors, since the temperature inside often reached between two hundred and three hundred degrees Fahrenheit.[53] Another common version, similar to the system employed by Strutt, featured "horses" that slid in and out of the closet. When fully extended out into the room for loading, a board at the back sealed off the closet and prevented the escape of hot air.[54]

In contrast to washing and ironing, drying received very little attention either in trade literature or by machine manufacturers looking for ways to innovate. Progressive laundry owners who seized each new improvement in washing and ironing machinery kept their old drying closets well into the twentieth century. The big innovation of the early twentieth century was an endless chain and set of hooks that traveled on a track through the closet, conserving heat and sparing workers.[55] Although rotary tumble dryers were available commercially in 1912, they were not commonly used in commercial laundries until the 1920s.[56] Laundry owners probably paid little attention to drying because the closets did not cause a bottleneck, they required little main-

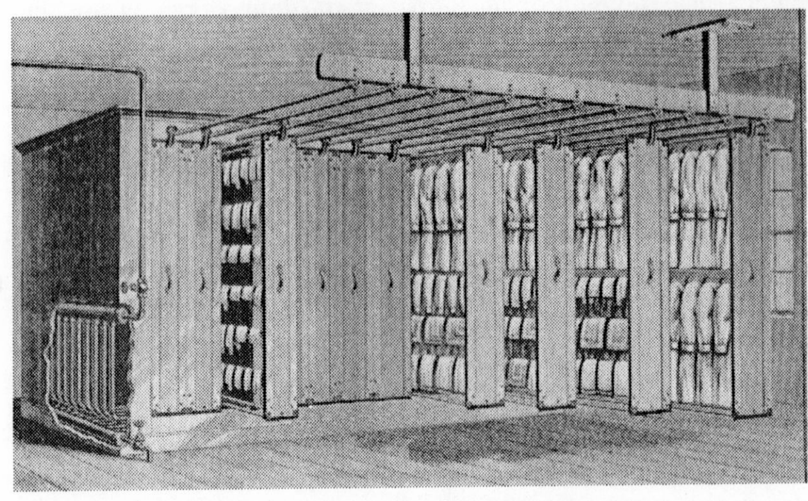

Fig. 12. Even in the most mechanized laundries, drying closets remained the preferred method of drying into the 1920s. From the Troy Laundry Machinery Company, *Illustrated Catalogue* (Chicago, 1907), 76. Courtesy of Hagley Museum and Library.

tenance or skill to operate, and they carried little risk of damaging clothing compared with other parts of the process.

From the standpoint of efficiency and protection against air pollution, drying rooms were also a huge improvement over simply hanging clothes, either indoors or out. Still, some consumers continued to cling to the notion that outdoor drying was preferable. To cater to these ideas, some laundries implied that goods were actually dried outside. In cities fouled by coal smoke or frequented by inclement weather, this was hardly a possibility. Still, small laundries played up their lack of a dryer as a virtue even if this meant workers dried clothes in the same tenement rooms in which they ate, slept, and used the water closet.[57] In more rural areas, a few large laundries did continue to maintain drying yards.[58]

Dried, starched, and redampened items were passed on to the ironing room. In the early years of the century, many laundries still used irons heated on specially designed coal-fired stoves. Often circular in shape, carrying tiers of heating irons of various weights, these stoves occupied the corner of the ironing room. Ironers relied on pressure as well as heat to remove wrinkles, so many irons were quite heavy, weighing from between four and twelve pounds depending on the purpose for which they were designed. They also had to be seasoned, usually with paraffin, so as to not rust or stick to the goods being ironed. Hand ironers who worked on fancy goods also used a variety of other tools. The frills and pleats of women's clothing required the use of specialized devices such as the goffering iron (a pokerlike implement) and fluting and crimping irons.

Inventors were always hard at work trying to make a better iron. In particular, they looked for ways to keep irons hot so that users would not continually have to travel back and forth to a stove. The surface temperature of stove-heated irons also varied depending on how long the iron had been heated and how the stove had been stoked (hence the ritual of spitting on the iron to test for heat). In the nineteenth century, manufacturers introduced various irons with internal heat sources. The most common contained burning charcoal or a metal "slug" that was heated in the stove and inserted into the iron.[59]

Gas irons were introduced at the end of the nineteenth century.[60] They had the advantages of staying hot and of providing some consistency. The first gas irons utilized a single hose that provided gas to the iron, feeding a flame just under the user's hand. Improperly adjusted, these devices produced an odor of uncombusted gas so strong that one could sometimes smell it coming in the door of the laundry. Continued use of these irons could cause various health problems. Thomas Oliver wrote: "In visiting large laundries I have been struck by the pale and anemic look of many of the women employed in the department where ironing by hand is carried on. Many of the women suffer from headaches and tender eyes consequent upon bending over the table and inhaling some of the coal gas which escapes from the tubes or jets and which is kept lit to maintain the temperature of the irons. Some of the symptoms are the result of high temperature and mild carbon monoxide poisoning."[61]

The addition of a second feed tube providing air helped curb the problem of uncombusted gas, but ironers continued to suffer when laundry owners improperly adjusted the gas feed or failed to replace cracked tubes and poor attachments that leaked gas.[62] These devices could also be disconcertingly noisy. Alice Linton said that in the British laundry where she worked, the "gas fired irons make a shrill screaming noise but one got used to it and the women would sing away above the noise of the irons and even hold a conversation."[63]

Because hand ironing was slow and good hand ironers were hard to find and expensive to pay, laundry owners yearned for ways to mechanize the process. However, designing machines to iron goods other than flatwork presented difficult technical challenges. Up until the start of the twentieth century, most ironing machines were variations on the mangle: heated drums between which goods were flattened. The earliest of these devices dates to about 1875 and had originally been designed to be used in collar factories. Compared with the bodies of shirts, the separate collars popular in the nineteenth century could be ironed relatively easily unless they required points or creasing. Machine operatives simply redampened them after starching and fed them through a small mangle with a single roller.[64]

These early devices were essentially variations on a flatwork ironer. The curves and contours of clothes, particularly shirts, presented a more complex design problem. Laundry machinery companies and laundry owners used both new devices and division of labor to solve the problem. In large shirt laundries, employers broke the process down into specializations: one worker ironed only cuffs, another only sleeves. Though some rotated between machines, the pressure of making a living on piece rates gave workers incentive to learn how to use a single machine efficiently.[65]

Most of these new ironing machines followed a similar set of design principles: a metal piece in the desired shape of the part to be ironed was heated internally by steam or a gas flame, the operator placed the collar or cuff on the machine, then stepped down on a foot treadle that opened and closed the machine around the portion to be ironed. Many treadles had simple mechanical linkages without hydraulics or counterweights to aid the operator, who had to apply her entire body weight to get them to open and close. Buttons, which got caught or broken in machines, further complicated the process.[66]

Like other heated devices, ironing machines posed a threat to the health of workers. One investigator wrote her small accidents off to inexperience, until her fellow workers told her otherwise: "At a sleeve-ironing machine, in another place I received some slight burn every day. And when I asked the girls if this was because I was 'green,' they replied that every one got burned at the machine all the time."[67]

Elizabeth Butler found that, despite the apparent advantages of ironing machines, laundry owners were sometimes wary of investing in them because they could not always be adjusted to changing fashions. Pittsburgh laundry owners endured a crisis when customers began demanding a "domestic" rather than a "polished" finish on their shirt fronts. One laundryman told her, "with considerable indignation," that the fickleness of consumers had forced him to buy two new $600 bosom presses.[68]

Goods finished in the various departments of the laundry were then delivered to the packing room, where they were re-sorted and compared to the record made on their initial receipt. Wet wash or wet dry work was put back into the basket or bag provided by the customer. Some laundries tried to convince these customers to buy special canvas bags because baskets did not stack neatly and had to be lined with oiled paper to protect the clothes. In a symbolic act, some customers sent their laundry in their galvanized iron wash tubs.[69] Some laundries also employed a seamstress or "mender," who repaired tears and sewed on buttons. Although laundries implied that this was an extra service to the customer, the seamstress also repaired damage done by the laundry itself.[70]

Finished goods were more elaborately packaged. At least some laundry owners believed that "a good housewife is painfully anxious about her linen and looks upon its cupboard as a treasure chest." Goods were therefore folded to show patterns or embroidery to best advantage. Shirts were pinned together or tied with a ribbon and wrapped in colored paper that "shows up the whiteness of the shirt."[71] Laundry supply houses could provide all manner of patented packing material: shirt boxes, collar stiffeners, strips of paper with the laundry's name emblazoned on them. This was ideal practice. In some markets, economics might dictate the simplest possible packaging: goods tied with a string and slipped back into the container in which they were delivered to the laundry. Packed into the laundry's wagon or shelved to await pickup, goods were then returned to the customer to begin the cycle again.

✖

Mechanization in the laundry industry was not characterized by dramatic changes. By and large, the process proceeded in a piecemeal fashion. The basic outlines of the process, as well as much of the machinery described in this chapter, would have remained familiar to those employed in the industry throughout the first half of the twentieth century. The sum of the changes, however, had profound implications. The work process in laundries bore less and less resemblance to its domestic counterpart. Workers from old-fashioned laundries found themselves out of a job or learning how to utilize unfamiliar machinery. They also found themselves facing new risks to life and limb. Laundry owners with enough capital to invest in new machines discovered they could operate laundries on an increasingly larger scale, and they found that machinery could be used to create excess capacity and to substitute for both brute strength and expensive skills. They also continually confronted all the ways the peculiar nature of the industry put limits on what could be achieved through these techniques: The raw material customers brought in was not always suited to machine processes, and increases in scale created new problems of control. Laundry owners would continue to struggle with these problems well into the twentieth century.

4

Women Workers and the Laundry Industry

IN 1906, THE ILLINOIS BUREAU OF LABOR STATISTICS ASKED LAUNDRY employers why they hired women for most laundry jobs. Some must have puzzled over a question that was rarely, if ever, discussed. In the end, respondents gave a variety of predictable answers, falling back on stereotypes proffered by employers in a wide variety of industries: "better adapted," "cheaper," "neater," "steadier," "more capable," "women's work," "women preferred."[1]

The language of the Illinois survey implies a greater degree of indeterminacy than either the history of the industry or its condition in 1906 would warrant. Both the British and Americans had established similar patterns of the sexual division of labor within laundries which, once in place, were rigidly observed. Perhaps investigators asked the question because a workforce that was nearly 70 percent female made laundries unusual among industrial workplaces.[2] Moreover, in other industries, jobs that were done by women in the home were not necessarily done by women in factories.[3] Despite occasional experiments with hiring men, the vast majority of laundries relied on women workers for everything but delivery, maintaining boilers, and minding washing machines.

Broad cultural ideas helped define what it meant to be a woman and a laundry worker, but the category was hardly monolithic. The older, independent ironer, the teenage "factory girl" operative, and the fastidious and polite clerical worker represented very different types, expectations, and ways of being a woman working in a laundry. Workers made distinctions within their own ranks on the basis of age, job description, and ethnicity or race. Employers also used these categories in assigning jobs and rewarding workers. Even reformers were not above employing stereotypes to explain certain behaviors or to engage their readers.

The relationship between these dual identities changed over time and with increasing organization and mechanization. The essence of that change revolved around the place of the older, married women who dominated the ranks of hand ironers. Their skills, work culture, and experience bridged the domestic with the industrial and connected smaller hand laundries with larger factory laundries. As long as mechanizing these women's work proved difficult, employers tolerated their independence and demands for high wages. However, mechanization offered an opportunity not just to save money but to hire workers who would better conform to employers' expectations of appropriate feminine behavior. Some middle-class reformers also welcomed more mechanized factory laundries as a means of removing laundry work for profit from domestic settings and disciplining or imposing order on workers.

Traditional Laundry Work

Small hand laundries ("cottage laundries" in Britain) were the form of commercial laundry work closest to home. They constituted one end of a historical and technological continuum, at the other end of which were highly mechanized factory laundries. Particularly before mechanization became widespread, workers–especially hand ironers–went back and forth between the two types of laundries, taking their skills and their work culture with them.[4]

Petty entrepreneurs often ran these smaller businesses out of their homes or in rented rooms. Basements were converted to washrooms, and wet clothing dripped onto the heads of ironers who toiled away in kitchens and living rooms. Families ate and slept surrounded by other people's dirty clothes and bedding.[5] One Chicago laundry that advertised itself as the establishment of a "common, sundried, grass-bleaching washerwoman" consisted of three tenement rooms. "The washing is done in the kitchen in tubs," the investigator noted. "The watercloset, the door of which was gone, opened out of kitchen. Clothes were dried upon lines overhead in the middle room in which were two beds and a dirty carpet on the floor."[6] Larger hand laundries occupied basements and first floor shops, but they often had the same improvised quality.

Even more so than steam laundries, hand laundries were the domain of women. Charles Booth noticed the difference in the washroom: "where machinery is in use, the washers are usually male, while, in hand laundries, men are only employed as dollymen, who punch or pound clothing in a large wooden tub with an instrument known as a dolly."[7] In other hand laundries, only women toiled over the tubs.

Hand laundry work was largely an occupation for adult, married women.[8] Without machinery to feed, there were no jobs for young girls. On one end of the process, immense physical strength and stamina were needed to scrub and lift endless loads of dripping linen from tub to tub. At the other end, ironing required both skill and a strong back as workers guided their heavy irons throughout a twelve-hour day.

Flexible work hours helped make employment in hand laundries appealing to married women. If a laundry worker was needed to tend a sick child or manage a household crisis, they could come to work late. Employers paid by the piece, allowing late workers to speed through their work and make up lost time. On the other hand, these employers also expected workers to stay late on Friday and Saturday nights until they had finished everything. In neighborhoods with large numbers of such laundries, it was not unusual to see women emerging at midnight or even later. Work was often seasonal. Many workers counted on periodic laundry jobs, to coincide with male unemployment.[9]

Although this form of laundry work seems to have been equally common on both sides of the Atlantic, British reformers examined it in greater detail and with a more critical eye.[10] They categorized hand and "cottage" laundries as a kind of sweated labor or homework.[11] Reformers found this type of work abhorrent for both its physical and moral effects on the worker and her family. As middle-class consumers, they questioned whether a trade that produced cleanliness as its primary product could be carried out under these conditions. In their reports, they emphasized the squalor of workplaces, the mixing of work and home life, and the straightened circumstances under which women were compelled to cling to such employment, despite its obvious shortcomings. One such woman told the Royal Commission on Labor that in the cottage laundry where she worked, "the washing was done in the kitchen, the ironing in the front room, and the drying also." Despite the fact the laundry was bitterly cold and filled with the stench of the "six or seven cats" and two dogs also living there, she stayed on–mainly because the laundry was next door to her own house.[12]

Laundresses also attracted other criticisms less typical of the debate over sweated labor. A wide variety of observers agreed that they had a propensity to drink, too often supported shiftless men, and were foul-mouthed and physically aggressive. One commentator neatly tied two of these complaints together: "In Acton [a neighborhood identified with large numbers of laundries] the women all drank but they worked; the men drank and did not work."[13]

Laundrymen liked to portray themselves as the hapless victims of these women. Until the mid-1890s, British trade journals often carried stock stories about the violent misdeeds of female laundry workers. Many of these stories mirrored the tale told in the August 15, 1887, issue of the *Laundry Journal,* which described the attempt of a "drunken laundress" to assault her male supervisor. He claimed she took after him with a hammer after he denied her requests for money.[14] In 1887, laundry owners used this stereotype to argue against including laundry workers under the Factory Acts. The editor of one trade journal opined that the "Factory Acts [are] intended to protect the young and the weak . . . laundry workers are, if young–as few are–by no means weak. They are, indeed, not only strong but they know it, and what is more do not hesitate to use their strength with relentless indifference to any consequences with which their actions may inflict on their employers."[15]

Because laundresses were reputed to be the supporters of ineffectual men, others ascribed their independent qualities to their lack of proper male supervision and the reversal of gender roles in the home. However, some commentators questioned the direction of causation. While admitting that many women went to work because their husbands were in one way or another genuinely incapacitated, these observers claimed that ne'er-do-well men attached themselves to laun-

Fig. 13. In hand laundries, washing was still done in sinks and clothing hung to dry on racks. The spaciousness of this early twentieth-century Boston area laundry belies the cramped conditions in which many of these businesses functioned. Currier Photo, ca. 1905, courtesy of the Library of Congress Photo Collections.

dresses knowing that such women, unlike shopgirls and many other operatives, could earn enough to support an unemployed husband. Employers and middle-class reformers tended to blame the men for forcing women to step into their shoes, frequently recycling comments such as "The laundry is the place . . . for women with bum husbands, sick, drunk, or lazy."[16]

These purported characteristics of laundresses were perhaps also emphasized in Britain because they built on a stereotype of the washerwoman already endemic in British culture.[17] Popular portrayals sometimes reduced the laundresses' traits to a pictorial image: not thin, small, and consumptive, as many of them must have been, but rather as full-figured, large-breasted women, towering over men—the mother figure ready to give succor or punishment at a whim.

In the 1893 report and successive reports, British inspectors suggested that the gradual transition from unmechanized, hand or "cottage" laundries marked not only a technological divide but also a transformation in the identity of the laundry worker. Inspectors welcomed mechanization and the imposition of factory discipline, believing these new types of workplaces would draw a clearer line between home and work, eliminating irregular work hours and workers' habits of eating and sleeping in workrooms. They lauded the disappearance of the married, hard-drinking, sewer-mouthed laundress as an industrial type, but they sometimes failed to acknowledge that the well-regulated "factory girl" who replaced her was often underpaid and underage.[18]

In 1907, Helen Bosanquet, who had campaigned hard for the extension of the Factory Acts to laundries, drew these parallels between mechanization and social control and voiced her own satisfaction in the changes in the workforce. "Many of the older set of workers object to the restrictions in a modern laundry," Bosanquet wrote, but she continued: "I think there is no doubt at all that the present generation of laundry workers is steadier, more sober, more efficient, and in every way more to be relied upon than the generation brought up in the old, unregulated laundries."[19]

In *Off the Track in London,* the urban chronicler George Sims vividly described his encounter with these juxtapositions during his wanderings through "Laundry-Land" in the first decade of the new century. Part of this landscape was the homes of laundresses who worked in their basements and hung their rooms with washing—a vanishing breed. "Farther away towards Latimer Road," wrote Sims, were the "great steam laundries employing a small army of young women, who at the dinner hour will turn out and make every street in the Dale a forest of white aprons."[20]

As steam laundries spread into neighborhoods once characterized by washerwomen and hand laundries, increasing numbers of women sought employment in them. In practice, the distinction between hand workers and mechanized laundry workers remained more ambiguous than people like Bosanquet implied. Women who had once stirred the coppers in hand laundries largely disappeared, surviving only as pitiful, aging figures retained by steam laundries to wash delicate items in the corner of the washroom.[21] However, before the years 1910-1920, many steam laundries still had only partially mechanized the ironing process. They took on large numbers of hand ironers to finish items of clothing that could not be put through the flatwork ironing machines.[22]

These women were the highest-paid workers inside most laundries. Ironers earned substantially more than shakers and machine operatives, who earned wages too low to survive on, by most estimates. The 1911 Senate study found female laundry employees in some American

Fig. 14. Large commercial laundries utilized both hand and machine ironing. G. Cadogan Rothery and H. O. Edmonds, *The Modern Laundry* (London, 1902), 351. Courtesy of Hagley Museum and Library.

cities earned as little as $6.00 a week. This set them well below the mark of $8.00 a week widely considered to be the minimum necessary for self-sufficiency. In contrast, some hand ironers earned as much as $15.00 each week.[23]

Laundry owners paid for what they and others called *skill*–a quality not often ascribed to other women workers in an industrial setting. Like the male workers designated as "skilled," ironers had knowledge and abilities gained through specialized training that was not immediately transferable to untrained workers.[24] Aside from the six- to twelve-pound flat irons, they employed a variety of implements such as crimpers and goffering irons. The ritualized testing of the iron with saliva had a real function for these women, who had to gauge the heat of the iron carefully–too hot and the goods would burn, too cold and time would be wasted by having to frequently return to the stove for another iron. The problems of the job were complicated by the fact that many items were finished with starch and sometimes other chemicals, all of which could scorch or change color if mishandled. Irons also had to be frequently reseasoned with a mixture of soap and paraffin to prevent rusting or stickiness. Stove-heated irons did not emit noxious fumes, but they added work. In one laundry that employed a hybrid scheme, an annual report described the shortcomings: "Each woman is allowed one jet, protected by a metal shield. Whenever she changes her iron she must walk around her ironing table, and often half way across the room to get a fresh iron. This takes both time and strength and it also unduly hastens the cooling of the iron."[25]

None of this differed significantly from what many women did at home. However, these ironers worked at a high speed, in rooms that could be overwhelmingly hot and humid. Shifts normally lasted ten to twelve hours, although during rush periods workers might remain in the laundry until late into the night. The most valued workers were those that were highly consistent. Ruined goods could not simply be tossed out as wastage. They already belonged to someone who would demand an explanation and compensation and would perhaps withdraw his or her business from the laundry.

Laundry owners continually complained about the shortage of good ironers and the high wages necessary to retain them. Valuable skills gave ironers leverage over their employers that operatives could not muster. Elizabeth Butler was told that ironers "were known as an unusually independent group, ready to leave in an instant if they fancied themselves offended, and ready to increase their value by bargaining, now with one employer and now with another."[26]

While in many industries the sexual division of labor worked to the disadvantage of women, the strict gender-specificity of laundry work helped female ironers hold on to high paying jobs. Gender-based boundaries around the learning process was one of the ways women retained control over skills. Whether in a kitchen or workroom, women learned to iron from other women. Just as males might learn the use of tools such as hammer and chisel, female ironers learned to use a variety of irons–tools identified with their gender. Machine operatives might also be said to have skills, but knowledge about the machines originated with male managers, machinery salesmen, and designers.

Despite ironing's identification with domesticity, many women learned to iron through formal or informal apprenticeships in a process that self-consciously mirrored the ways men learned other jobs.[27] Some paid for their training, from either a skilled worker or forewoman in a shop. In a few laundries, this apprenticeship took a strict form. One laundry owner told a British parliamentary commission that all of the ironers apprenticed in his laundry were assigned to an ironer who was paid two shillings a week for teaching. The teacher also received what a girl earned while under her tutelage, and she made her own arrangements as to paying her student some portion of this amount. Other apprenticeships were longer, with the teacher receiving some portion of the student's pay for three months or more.[28]

Paying for such training was no guarantee of success. In 1912, a Mrs. Harrison of New York's West Side was forced by family circumstances to seek a job. As the social worker who described her case stated, Mrs. Harrison had, at the age of thirty-seven, taken one of the lowest-paid laundry jobs and was "glad to shake out sheets in the laundry next door to her home for $4.50 a week."[29] Mrs. Harrison apparently needed more money, probably to help feed her family, and the Charity Organization Society arranged to give her three weeks' training as an ironer so that she might be able to do the more skilled and better paid work.

Three weeks was a minimum training period in this laundry.[30] Mrs. Harrison, after two weeks' trial, decided that she was "too old to learn," and fell back on shaking at her old wage.[31]

With male workers, skill correlated not only with higher pay but also with independence, status, and respect both inside and outside of the workplace. In contrast, a "skilled" female worker was in many ways an inherently conflicted role in late-nineteenth-century Anglo-American culture. Cultural understandings of appropriate gender roles for women and skilled workers sometimes ran in direct contradiction to each other (in many ways, they still do). How is she supposed to act? Should she be manly? Can she drink and swear and stand up to the boss and still retain her social standing? Should she have craft pride? Can she take that pride home with her at night as she prepares to work the "second shift"? How is a manager to deal with such workers? Man to man, man to woman? middle class to working class?

In the end, individual women worked out various answers to these questions. They were aided, though, by a work culture that emphasized pride and defiance in interaction with employers, hierarchy among women on the shopfloor based on skill, and strategic accommodation of the conflicting demands of home and work.

Much to the chagrin of reformers and many employers, ironers and older women seem to have been the ringleaders in importing the work culture of hand laundries into steam laundries (including beer drinking and willful disregard of their employer's demands for punctuality and other aspects of factory discipline). Employers alternatively tolerated, exploited, and tried to eliminate these attitudes and behaviors.

In some laundries, the problem of skilled workers ignoring schedules was so pressing that employers resorted to a system of bonuses as an incentive. A gratuity of one shilling a week was credited to those ironers who were both regular and punctual in coming to work. Employers began compounding the payment in November and paid it out in a lump sum to those who were still employed at the company in August of the following year.[32]

If employers disciplined these workers at all, it was usually through a system of fines. The Women's Industrial Council found that "in some laundries there are no fines, the workers being simply dismissed if they fail to obey the rules, but in others 6 d. may be the fine for scorching an ironing board. Ironers are often fined for scorched work and for coming downstairs in couples."[33] The same investigation found that other employers were not above exploiting these women's independence by illegally treating them like subcontractors. Ironers were charged for materials and rental on tools: "It is complained that one schilling in the pound is deducted for the use of gas irons, etc., even when not used during the whole week. In some laundries 1 d. a week, in others 6 d. a week is charged for the use of the ironing boards."[34] One way to control soldiering on company time was paying through piece rates–a practice that became quite widespread after the turn of the century.

Laundresses also clung to other preindustrial work practices. In less mechanized, more traditional laundries, workers were often paid partly in buckets of beer, used to quench a thirst developed doing hot, heavy work.[35] With mechanization, some employers made an effort to eliminate alcohol from the workplace. Other employers still contributed for buckets or bottles, or they simply turned a blind eye: accommodation kept the peace and relieved them of the responsibility of providing fresh, cold water.[36] Those that opposed workplace drinking rightly recognized it as a preindustrial practice. They saw the link between drunkenness and accidents with machines and the way the presence of alcohol symbolically undermined factory discipline.

Many older women drank anyway. One American investigator got this explanation from the "boss starcher" delivered with what the investigator described as "quiet elegance": "They [the workers] did not think of such a thing as drinking beer behind the boss's back, but they just didn't want him to know."[37] In many laundries the division was generational. The same investigator visited another laundry, in which the tap water was so foul tasting as to be undrinkable. The younger women drank bottled soda water while the older women drank beer.[38]

Skilled workers also demanded deference from their fellow employees. This came out particularly at the lunch hour. In some laundries the high wage earners sat apart from the low wage earners. Clark and Wyatt described the scene in another laundry where "the high wage earners, though they often treated the $5 girls to stray sardines, cake, etc., were in the habit of sending young girls to the delicatessen shop to get their lunches and also to the saloon for beer. . . . For this service the girl would get 10 cents a week from each of the women she did errands for."[39]

When many of these investigators observed hand ironers, the latter were already a dying breed. Elizabeth Butler described this new mechanized workplace: "As you enter the ironing room on an upper floor of a laundry, you hear a rush of belting, the irregular sound of reversible rolls, and the sharp quick clamp of a metal press." All the work was divided into minute proportions. "No girl," said Butler, "irons a whole article. Instead, she irons a sleeve, or a cuff, or a yoke, or perhaps one side of a collar."[40]

As in many other industries, mechanization meant the fractionalization of jobs, the replacement of skilled workers with unskilled workers, and the gradual disappearance of a preindustrial work culture. As this pattern played itself out in laundries, however, men did not have a significant role—women replaced women. In particular, young, single operatives replaced middle-aged, skilled workers, and the highest paid jobs shifted from women doing craft work (ironing) to those doing clerical work (checking and marking).

Mechanization was a gradual process. As the technology became available, it was snatched up by some laundries while others mechanized only partially or not at all. The reasons for the piecemeal nature of the process are complex. The technology itself was not without problems. A worker who ruined work could easily be fired and replaced, but a faulty machine was costly to repair or replace. Moreover, ironing machines were designed for specific kinds of work, each suitable only for clothing of a particular design: round collars, pointed collars, flat pleats, puffed pleats. When fashion changed, a skilled worker simply learned to adjust. For the mechanized ironing room, a change in fashion might require a new machine.

Like other small and medium-sized businesses, many laundries ran on very narrow margins of profit. In the period around 1910, a machine for a single part of the process could cost hundreds of dollars. Some business owners simply could not afford this kind of investment, so part of the lag in mechanization was caused by these small owners waiting for machines to appear on the secondhand market.

Employers' decisions to mechanize fundamentally changed not only the way that work was done but also the nature of the workforce and the social relations between workers within the laundry. Though all processes in the laundry were subject to mechanization, introducing machines that eased processes requiring a great deal of physical strength had a different effect than introducing machines that replaced skilled workers and divided tasks. Cecil Roberts discovered the difference when he installed a new set of ironing machines before the First World War. He eventually concluded, that it was "a mistake to try to turn an experienced hand ironer into a machine operator–it was better to train young girls who had no previous experience."[41]

The incentives and consequences of mechanizing skilled labor processes were complex. Further, it is not clear whether the consequences laundry owners intended when installing machinery were always consonant with the results achieved. Some evidence suggested machinery did not dramatically lower costs. The 1920 U.S. Census of Power Laundries revealed that over the previous ten years, mechanization had scarcely decreased the percentage that wages figured in overall costs. In 1909, wages had constituted 50.6 percent of the cost of running a laundry. By 1919, that figure had only decreased slightly, to 48.9 percent.[42] Moreover, the cost of buying new machinery negated much of the savings on labor costs. Eventually laundry owners did save money on wages, but the motivations for mechanization were more complicated than this simple bit of economics. Many laundry owners did not have the business savvy to figure out the precise economics of mechanization for their laundries. It was enough that they believed machinery would make them money and help with labor problems.

The most obvious benefit of machinery was the speed with which work could be done. The ironing room had long been the bottleneck of the laundry. Mr. R. W. Barber took the editor of the *Laundry Journal* on a tour of Spier and Pond's, one of the largest of London's laundries. "Before leaving," said the editor, "I had a look at the finished work of which Mr. Barber is justly proud. 'We get a lot of this class of work' he said fondly fingering the real lace on a lady's robe de nuit. 'Of course we charge proper prices–this would be half a crown at least–but it's fearfully trying. You can't get enough skilled hands, and it blocks all

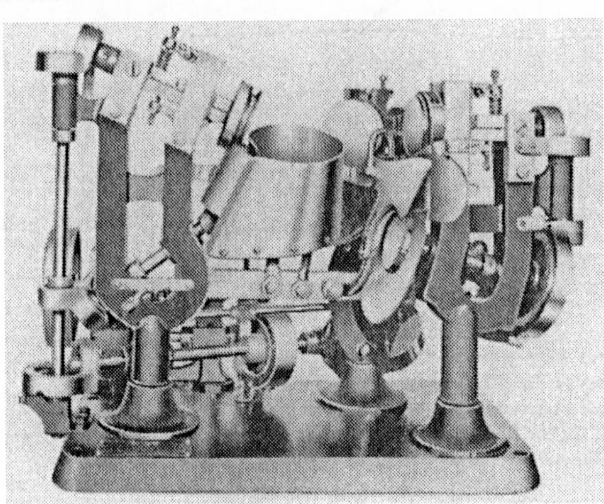

Fig. 15. The hand iron was a fundamentally different type of technology from ironing machines such as the collar shaper. Cheltenham ironing stoves from W. Summerscales and Sons, Ltd., *Catalogue* (1897), 78. Courtesy of Science & Society Picture Library. Collar shaper from Troy Laundry Machinery Company, Catalogue, 1915, n.p. Courtesy of Hagley Museum and Library.

the other work. No–I like collars and cuffs best–things we can do by machine. Some day, perhaps my ideal [of full mechanization] will be realized, but–well look at that long room' pointing down a vista of hand ironers 'there's plenty of work for hand ironers yet, isn't there.'"[43]

Laundry owners also found machinery to be a solution to the perennial problem of unexpected excess work. Most laundries could not control the number of bundles coming in from week to week, and they could only roughly predict it. Before the late 1890s, they had most often relied on overtime to handle excesses. Ironers in particular often stayed until very late on Thursday and Friday nights. Ten-hour laws and restrictions on nightwork for women effectively put an end to these practices, especially in Britain. Machinery offered another means to create extra capacity.

Machinery also allowed managers to move workers around so that the sudden loss of a skilled worker would not cause a crisis in one part of the process. The expertise (supposedly) was in the machine. More subtly, ironing machines solved what might now euphemistically be called "problems with personnel." Managers no longer had to deal with the endless struggle of hiring suitably skilled workers who knew the value of their abilities and who often had the age and self-confidence to shape their work lives to their own needs. After buying machines, owners could replace these women with tractable, unskilled teenagers and single women.

Characteristically, Americans were more enthusiastic about mechanization than their British counterparts. British reformers, for instance, worried about the health consequences of the rising tide of foreign-made machinery. One noted that "new and ingenious inventions constantly appear, many of which come from America, whence a considerable amount of this machinery is imported."[44] In Britain, the cultural resonance of the traditional laundress's image both prolonged older practices and added incentive to the process of mechanization. The feisty, independent washerwoman was a stock figure in British popular culture, whose last refuge became the ironing table of the steam laundry. She provoked strong feelings–from both her admirers and her detractors–that had no equivalent in American culture. British laundry owners lamented over and over that they would have liked to have kept her on if only she would work for less, produce more, quit less often, and generally behave in a becoming and subservient manner.[45]

The editors of the *Laundry Journal* reported from census figures that the age of laundry employees in London had dropped because of mechanization: 21 percent of the workforce had been under the age of twenty-five in 1891, a figure that rose to 34 percent by 1901.[46] The next two decades witnessed even more substantial changes. Though similar figures are not available for American laundries in the same period, anecdotal evidence suggests a similar pattern.[47] In the United States, mechanization was also marked by an increased influx of African American workers into laundry jobs as white women looked for better paid employment in other industries.

As the percentage of operatives in laundries rose, wages remained stagnant compared to the cost of living. A 1918 investigation into New York laundries judged $11.70 a week to be the minimum living wage for single women on their own. The report stated that one-quarter of the workers earned $6.00 or less, one-half $8.00 or less, and four-fifths $10.00 or less. Only about 20 percent were earning enough to live on.[48]

The immediate effect of mechanization was a shift in the shopfloor hierarchy among workers. In most laundries, machine operatives did not rise to the top of either the social hierarchy or the wage scale. While employers acknowledged that it was important for machine operatives to have "skill," in the sense of experience and ability, they exercised more control over its acquisition. Machine operators did not learn the job in their mothers' kitchens. Youth and strength came to matter more than experience and know-how.

It would be misleading, however, to suggest that skill was the only way in which workers distinguished among themselves or were differentiated by employers and reformers. Race, ethnicity, class, age, and demeanor all affected hiring, assignment to jobs, and relations among workers and between workers and managers. The workforce of most British laundries was racially and culturally homogeneous. In contrast, distinctions on the basis of race and ethnicity further complicated the situation in American laundries. Many American employers believed that there was some connection between ethnicity, race, and skill. They preferred Northern or Eastern Europeans–Irish, German, Polish, and English–over Southern Europeans and African Americans. Before about 1915, most laundry workers in the Northeastern and Midwestern United States were of Northern European origin. In spite of their large representation in the clothing trades, Italians and Russian Jews were a barely noticeable presence in larger laundries.[49]

By the same token, American laundry owners were extremely unwilling to hire African American women to work in laundries before the 1920s, despite their prevalence as washerwomen in domestic service. The Charity Organization Society in New York, which trained both African American and white women to do laundry work, did not even bother trying to place the black women in commercial laundries. Laundry owners offered few explanations, although one can speculate that worries about customer preferences and prejudices, along with prejudice among other employees, all played a role. Laundry owners bought into stereotypes about the cleanliness and reliability of African American workers. They also worried what their customers would think about such persons handling their private things.[50]

Employers' prejudices about race and ethnicity also shaped placement for certain tasks within the laundry. Each manager, foreman, or forewoman had their own notions of what group was most suited to what job. After the installation of ironing machines in one Pittsburgh laundry, the foreman told Elizabeth Butler: "No American can stand this. We have to use Hungarians or other foreigners." (Butler noted that other laundries found Americans perfectly suitable for the same machines.)[51] Wage differentials also reflected these prejudices. African American and Italian workers often got paid less for doing the same jobs as their Northern European counterparts.[52]

Employers who ran what they liked to think of as "high-class" laundries were also willing to pay more to women with certain class or racial characteristics, regardless of the quality of work they were able to turn out. A white, native-born woman who had been forced back into the labor market by her husband's desertion told Senate investigators that the "better class of laundries have [a] good class of girls, and the wages are better and the work easier than in many other kinds of work."[53] Buried in her statement was the implication that being part of that "good class of girls" meant being white, Northern European if not native-born, and having at least some education. Native-born women seem to have been especially loath to mix with other ethnic groups in the workplace.

For British employers, a "better class of girls" literally meant women drawn from the artisan and lower middle classes, whose "education, taste, and refinement" might make them more malleable and conscientious employees than their lower-working-class sisters.[54] "Take the ironing room," argued a laundryman in the pages of a magazine for working women, "who is more fitted for this work than a lady?" Among his reasons, he stated that a woman of a higher class would have a sense of dedication to the job. He believed that she would not "suddenly throw up her work in the middle of the week, and thus inflict a serious injury on her employer, perhaps out of a spirit of pure restlessness, or because she dislikes some fellow worker. Nothing of all of this would an educated woman do."[55] Nor, under most circumstances, would she accept the combination of mediocre wages, social stigma, and hard labor attached to laundry work. His plea probably attracted few of the type of women he sought.

Employers also distinguished between women on the basis of demeanor. In hiring operatives and ironers, they often passed over women whose behavior conformed with middle-class standards of femininity in favor of those that seemed tough enough to withstand the work. Katherine Anthony objected to the way that "with the women, as with the men, it is often the 'pushing sort' that gets the job rather than the industrious steady worker." Anthony held up as an example two women: Mrs. Rudden, "who hadn't a 'lazy bone in her body,'" but did have "a shy, apologetic air which was a distinct disadvantage in looking for work. Applying at one laundry after the other, she was always among those who were turned away." On the other hand, young Mrs. Snyder, with her "hustling manner and her engaging smile used to declare that 'she could always get taken on, even if there was 20 turned away.'"[56]

In the highly mechanized laundries Elizabeth Butler visited, "in general, wages are paid to the girl, not the position. The specific amount depends on the appearance of the applicant, her quickness, regularity, and her ability to bargain."[57] Age also seems to have been a factor. The same investigation found that, overall, women over the age of thirty earned more for the same work in every job category than their younger counterparts. Very young operatives were the worst victims of this system. Employers often managed to squeeze several months of work out of them on "learner's wages." This phenomenon significantly sets laundries apart from other industries, in which piecework rates and deskilling ensured that youthful speed and endurance were the only guarantors of higher wages.

Compared to the divisions described above, it is difficult to say conclusively what difference the greater homogeneity of British laundries made in shopfloor culture. In her book, *English Laundresses,* Patricia Malcolmson says that many English laundry workers described their places of employment as "homey." Their descriptions in autobiographies and reformers' reports suggest that they were not referring to some quality of *Gemütlichkeit* but rather the replication of work relations in kitchens and other sites of women's work. In a sense, British employers replicated the paternalistic conventions of the Victorian household more conspicuously than their American counterparts. For instance, British trade journals regularly carried news of the parties given by employers for their staffs. An apparently marvelous time that all members of the staff had at the coming-of-age party of Mr. Norman Raintree, the son of the head of one large laundry firm, was described in glowing detail for the readers of the *Laundry Journal.*[58]

One of the most explicit expressions of the paternalistic atmosphere of British laundries were the annual company parties. Universally referred to in the industry as "bean feasts," these celebrations provided a forum in which employers, managers, and wage labor mixed socially. These festivities were apparently ubiquitous among larger laundries. In the January 19, 1907, issue of the *Laundry Journal,* a number of these parties were described, including the Latchford Steam Laundry Social and the Hygienic Laundry Social.[59] When many British laundries announced that they would cancel these annual outings on account of Edward VII's coronation, an anonymous writer for one of the trade journals warned: "It is not every employer who realizes the tremendous business value of these bean feasts. The amount of good feeling engendered between employer and employed will make the wheels run more smoothly all through the year." He added, with a bit of Victorian prudery, "Properly managed there is not the faintest reason why these outings should degenerate into orgy."[60]

"Homey" was not a word that American laundry workers seemed to use to describe their workplaces. Undoubtedly, some American employers treated their employees to picnics or offered both welcome and unwelcome gestures of paternalism, but these practices were not an entrenched part of the culture of the industry as a whole. In these and other laundries, ethnicity and later, race, created a cultural divide. Employers could not imagine these workers as part of their family in any capacity. The feeling was probably heartily reciprocated.

One of the most concrete manifestations of the effects of heterogeneity versus homogeneity in the workforce lies in the gender differences of management. In British laundries the managers were quite frequently women, while most American managers were men. Unlike American forewomen who supervised a single department and often worked alongside other workers, the British manageress often managed all parts of the laundry except the washhouse and the steam plant, where the engineer presided.[61] This difference illuminates the nature of a culturally homogeneous workplace: The manageress derived her authority from her class status.[62] Whereas in the polyglot, culturally diverse atmosphere of the American laundry, the subtleties of accent and deportment were lost on many workers. American managers were most often men who derived their authority from the more visible, easily translatable category of gender.

In Britain, female managers were hired because their presence in laundry made sense for business. Not only were they cheap, receiving far less pay than their male counterparts, but they could also negotiate with female clients and workers in their own terms. In certain circumstances and in certain laundries, they could even be expected to pitch in during extremely busy periods, doing the women's work that was considered to be below male managers. It is likely that their management style was also gender-specific, depending on negotiation and quiet pressure. In August 1890, the *Laundry Journal* asked a Miss Prince of the New Era Laundry to describe her "secret to success" as a manageress. "First," she replied, "keep your eyes open, and secondly, be firm, I don't mean to say be unkind–there's no need for that, but I always say straight out what I mean."[63] The issue of being "kind" was less likely to be part of the considerations of male managers, not because they were inherently more unkind than the manageress but because they often seemed to see women workers as the "other": a teeming mass of interchangeable components necessary to make the plant as a whole function efficiently.

Female managers could also be a guard against the sexual harassment and sexual favoritism that characterized some American laundries. Dorothy Richardson described such a favorite, nicknamed "the Queen," who worked in the New York laundry where Richardson was employed. Big and blond, "a voluptuous beauty," "the Queen" was a favorite of the foreman, who left her to "boss the job" when he wasn't around, much to the resentment of the other workers.[64]

Despite their positions of authority, managers (male and female) were, in the end, only employees themselves. Even those with good intentions were only in the laundry to enforce the policies of the owner. They could make few changes in regards to wages, hours, or working conditions.

Women managers had other problems with male peers and supervisors. Managers or not, they were still women, and their presence in the masculine business culture of "laundrymen" made some of their male counterparts nervous. A Mr. Roberts, asked to give a toast to the "ladies" at the 1899 dinner of the National Laundry Trade Protective Association, masked his hostility with backhanded compliments, reportedly saying that he "always looked upon laundry ladies as so desperately serious and businesslike that he hardly knew whether he ought to tackle the subject in a serious or jocular style. At all events, in the laundry trade, ladies were a *sine qua non*, for the trade could hardly get on without them, and it was quite a question who was pre-eminent, laundrymen or ladies."[65]

Twenty years later, when women managers were excluded from sitting on the laundry trade board, laundry owners took a softer tack. Hugh Trenchard, a prominent laundry owner, made a "frank confession," saying: "If I have trouble with the workers I lock myself up and leave my manageress to deal with it." Perhaps owners thought that these women could better disarm this threat to their industry by calming worker demands and charming government representatives, for the author of the article suggested that "any one of us could mention a couple of dozen women at least who, for business capacity or tact, could give almost any man on the Federation council 'two stones and a beating.'"[66]

Laundry owners disagreed about which side of the delicate border between middle and working class from which the "manageress" should come. Hugh Trenchard stated that he liked "a woman who has *practically* risen from the ranks–a small tradesman's daughter or something of that sort. . . . A lady isn't much use–she hasn't the physical strength and hasn't been used to the long hours in early life." He went on to qualify his remarks further: "And yet–to use a horsey phrase–you want a little more blood than you get in the rougher hands."[67] Others recruited straight from the lower middle class, gathering up what one laundryman indelicately called "waste girls."[68] More commonly known as "redundant women," these were middle-class women who were forced to earn a living usually because they had failed to marry owing to the demographic shortages of suitable men.[69] The laundrymen and some columnists for women's publications argued that laundry management provided a higher paying alternative for women who might otherwise become governesses or teachers or nurses. A few gentlewomen even ran schools to teach these women. These establishments were sometimes described in the trade literature. A Mrs. Leith Wright, for example, ran a "Cottage Laundry School" for "lady pupils" to teach laundry work.[70]

⁂

Gender meanings, gender relationships, and gender identities permeated the work of laundry. They influenced who did what job, how they behaved when doing that job, and how much they got paid. Gender requirements, however, were neither fixed nor monolithic. Workers and employers distinguished between women most obviously through hierarchies of age and skill. As mechanization reached the most skilled parts of the process, those hierarchies also shifted. Ethnicity, race, and class also complicated identities and relationships, creating significant differences between British and American workplaces.

Unionization and Regulation, 1888–1914

THE LAUNDRY INDUSTRY EMERGED AMIDST AN EXTRAORDINARY period of industrial reform and working-class agitation. Those trade union activists and middle-class reformers dedicated to ameliorating the ills of women's industrial employment seized on laundries as a quintessential "woman's industry." Previously, laundry workers had depended primarily on informal methods to improve their circumstances–individually bargaining with employers, shifting jobs, soldiering–rarely forming organizations or striking. At the end of the nineteenth century, reformers and organizers on both sides of the Atlantic began to prescribe a combination of unions and legislation as a more effective means of reshaping the industry.

Organizations representing female reform initially preferred collective bargaining as the method to improve the working woman's lot. The other available option–protective legislation–restricted women's freedom to choose the terms of their own work. However, unionism, in all its varying forms, proved a continual disappointment. In the United States, mixed-sex, American Federation of Labor (AFL) sanctioned locals could draw on the federation's financial resources and expertise, but in the bitter aftermath of a series of strikes, women workers learned their male compatriots continually betrayed them at the bargaining table. In Britain, single-sex unions, formed to avoid this conflict of interest, were plagued by small and transient memberships. Even in their best years, both types of laundry unions took in only a tiny percentage of potential members in either country.

First in Britain and then in the United States, state intervention offered another viable option. Between 1891 and 1907, British reformers campaigned for the full inclusion of the laundries under the Factory Acts. They did not expect to mandate higher wages but instead hoped to limit overtime and improve the conditions in which women and underage workers toiled. In the mode of the day, investigation preceded action, generating a body of research and a flurry of articles in socialist and liberal journals. Much to the chagrin of many laundry owners, female laundry workers were singled out as public exemplars of the capitalist exploitation of female labor.

The vanguard of American labor legislation lagged behind its British counterpart, and impatient American reformers turned to Britain for tools and inspiration.[1] However, the nature of the American legal system, which vested power to legislate in the states and left those laws open to constitutional challenge, continually hindered their energetic efforts.[2]

Employers recognized that labor unrest and state regulation were not just a sideshow to the business of creating an industry. Nevertheless, they disagreed about the potential impact of these factors. Because laundries varied so much in size and degree of mechanization, different businesses would be affected in different ways. Laundry owners realized regulation would cost money and hinder efficiency for all businesses, but it could potentially harm small laundries much more than their larger competitors. These "small fish," as the British called them, lacked the capital necessary to make expensive modifications to machines and workrooms required by inspectors. They also relied more heavily on the overtime and nightwork reformers hoped to outlaw. Unionization could be a double-edged sword, increasing costs and lowering productivity but providing an additional means through which employers could discipline "unruly" employees. Finally, even if moderate regulation and union agreements might help winnow competition, laundry owners feared harsher measures would result in higher prices, driving customers back to the washtub or unregulated washerwoman.

Laundry owners and writers for trade journals had difficulty disaggregating the effects of reform measures from other factors transforming the industry. They often took the easy way out, relying on convictions commonly held in the business community rather than on hard data. These beliefs differed on either side of the Atlantic. The public pronouncements of British laundry owners and trade journals often favored some type of organization and regulation as a means of stabilizing the industry and rewarding "good" employers.[3] Meanwhile, a pervasive antiunion, antiregulatory ideology among American employers kept them from seriously considering this arrangement until the 1930s.[4]

Laundresses and Members of Parliament

In early 1889, Mrs. Barber, an upholsterer by trade and member of the Union of Chelsea Seamstresses, set about organizing a laundresses' union among her London neighbors.[5] Unlike Barber's sister upholsterers and seamstresses, laundresses were not known for their interest in unions. They belonged to the legion of female and semiskilled workers previously considered unorganizable by the male-dominated leadership of the Trades Union Congress (Britain's principal labor federation) and women's reform groups.

However, the late 1880s were a time of high hopes for workers outside the traditional purview of English trades unions. Under the banner of "New Unionism," strikes by dockworkers and female matchmakers sent a message that both unskilled and female workers could be successfully organized.[6] Mrs. Barber and the hundreds of London-area laundresses who flocked to the new union took inspiration from these events. Soon thereafter, the Amalgamated Society of Laundresses boasted 3,000 members and eight branches in London and Brighton.[7]

Like other women's unions, the Amalgamated Society of Laundresses drew on the resources of the Women's Trade Union League (WTUL), an organization of women workers and their middle- and upper-class "allies." From her house on fashionable Sloane Square, Lady Emilia Dilke helped direct the WTUL's efforts. She found the laundresses additional organizers, including her secretary, May Abraham, and turned up in churches and union halls to give speeches (and probably money) to the local branches.[8] Dilke also openly supported the laundresses' campaign to convince members of Parliament that their places of employment should be included under the Factory Acts.

This marked a radical departure in WTUL policy, as members had long resisted supporting protective legislation, arguing it was too often a ploy used by male trade unionists to rob women of their jobs.[9]

At the Trades Union Congress of 1890, representatives of women's unions came to ask the male delegates to endorse an extension of the Factory Acts. The men gave them an enthusiastic welcome, much to the women's surprise. When her turn came to address the assembly, the laundresses' representative vividly evoked the horrors of the trade. She told the audience that laundresses stood "ankle deep in water for fifteen or sixteen hours a day" under a curtain of dripping clothes hanging over their heads. As the assembly nodded in sympathy, she moved that the Trades Union Congress resolve that "no amendment of the Factory and Workshop Act would be satisfactory which did not include the extension of that Act to laundries." She later reported to the WTUL that "the delegates voted unanimously in its favor, and it was carried amidst applause."[10] The laundresses' cause offered the men an opportunity to support regulation without seeming to have an ulterior motive and without affecting their own economic opportunities.[11] Had the request been from female typesetters gradually moving into "male" jobs, the outcome might have been quite different.

Organization and solicitation of male trade unionists' blessings garnered astonishing results. One Sunday in June 1891, a huge crowd of twenty to thirty thousand people gathered in Hyde Park to demonstrate the labor movement's commitment to the laundry amendment.[12] Although such mass demonstrations had long been used by male trade unionists, such a gathering on behalf of women workers was unique.[13] Anxious to avoid conflict with the laundresses and their supporters, male labor leaders emphasized the familial and class connections between laundresses and male workers in their remarks to the crowd. One speaker reminded his audience that "the majority of these women were widows of respectable mechanics, and they had families to support." Two of the New Unionism's most prominent figures, Tom Mann and John Burns, found a place on the speaker's platform. Burns established his connection with the crowd by declaring that, as the London *Times* later reported, he "had the honor to be the son of a washerwoman and would rather be the possessor of that lineage than be the descendant of any baccarat-playing scoundrel."[14]

The rhetoric from the platform also reflected the still largely preindustrial character of the trade, in reality and in the cultural imagination. The Amalgamated Society took in both washerwomen and the employees of the largest factory laundries. Between the demonstrators on the lawn sat a van in which the women displayed "the implements of their trade," while two busily demonstrated their ironing skills–a performance echoing male trade unionists' similar displays of tools in public parades and demonstrations.[15]

The Amalgamated Society and their allies also directly lobbied politicians, stretching Victorian gender conventions in a way that attracted the bemused attention of the press. A deputation consisting of representatives of the laundresses, the London Trades Council, and the National Laundry Association (a trade group) went to lobby the Home Secretary, Henry Mathews.[16] In reality, the men probably did much of the talking, but *Punch* portrayed an encounter between Amazonian laundresses and a small, cowering Mathews. In the caption Mathews explains: "Can't see you now, I'm washing–myself."[17]

As the interested parties took sides, it became apparent that this issue evoked divisions more complex than capitalism versus labor. Not all employers opposed regulation or unionization. In 1890, the *Laundry Journal* went so far as to run a notice soliciting donations (to be sent to Lady Dilke's house) to help union representatives attend the Trades Union Congress conference.[18] "Onward!" the editor declared in October 1890. "If the proper spirit is shown, the labour unions and the laundry proprietors' Association ought to be able to meet and arrange all differences without friction or fear of fighting."[19]

By the same token, the decision to embrace regulation was not universal among laundresses or middle-class feminists. Laissez-faire liberals, including an older generation of female reformers, fought hard against the act.[20] At least some workers also feared displacement by men and machines. The *Laundry Journal* noted a rival "deputation of laundresses" also visited the cowering Home Secretary Andrews, proclaiming that, should the act be passed, "a large number of hands engaged in washing and laundry work would be thrown out of employment."[21]

290 PUNCH, OR THE LONDON CHARIVARI. [JUNE 20, 1891.

THE NEW TALE OF A TUB; OR, THE NOT-AT-HOME SECRETARY AND THE LAUNDRESSES.

"CAN'T SEE YOU NOW, I'M WASHING–MYSELF."

Fig. 16. In this *Punch* cartoon, Amazonian laundresses lobby for protection under the Factory Acts, coming to the door of Home Secretary Henry Mathews. From *Punch* (10 June 1891), 290.

Even if everyone in the trade could have agreed on the desirability of regulation, writing an actual amendment posed a daunting problem for lawmakers. Laundries varied so much in size and degree of mechanization that it seemed nearly impossible to devise the kinds of simple, uniform rules that governed industries like textile manufacture and mining.

On the Friday following the Hyde Park demonstration, member of Parliament David Randell introduced a motion amending the Factory Acts to include laundries. Randell and several other members laid out the arguments for inclusion: it was hypocritical to apply the acts to women workers in some industries but not in others; the amendment was supported by a large number of laundresses, as evidenced by petitions allegedly signed by 65,939 out of the 67,506 women canvassed; and many employers also welcomed regulation. Another member, Mr. Cunninghame Graham, made a more direct appeal to the members' sense of paternalism and noblesse oblige. He compared the long hours worked by laundresses to "white slavery" and told Parliament, "It is easy to make light of demonstrations in Hyde Park but I think any unprejudiced Member who had seen the component items of the laundrywomen's demonstration–the forlorn, draggle-tailed creatures who attended it, the impossible rusty crape bonnets and dejected looks–could not possibly have failed to be moved by it."[22]

Speaking for the conservatives, Mathews responded that regulation of hours would inconvenience consumers; that laundresses did not really want to be regulated; that the amendment was a ploy on the part of large laundry owners to put their small competitors out of business. Then, addressing himself to the Irish members of Parliament, he produced his trump card: the amendment, as written, would include laundries run by convents and other charitable organizations.[23] Such institutions often relied on revenue from their laundries. To operate these laundries, they depended on the labor of orphans, inmates, and other unfortunates–frequently performed under conditions that would not pass the scrutiny of a factory inspection. Exempting these "charity laundries" was not a politically viable option because laundry owners protested that such measures would give unfair competitive advantage to the Church.[24] Not enough votes could be garnered without Irish support, and the laundry amendment failed.[25]

The WTUL and its allies bound up their wounds and considered strategy, hoping to reintroduce the laundry amendment in 1895. The election of Gladstone's Liberal government boded well. True to expectations, the new home secretary, Herbert Asquith, sympathized with the cause. Soon after his appointment he organized a Royal Commission on Labor.[26] He also appointed May Abraham as one of the first two women factory inspectors and sent her to gather information on the conditions in laundries across England. Soon thereafter, Abraham and other women civil servants testified before the Royal Commission, vividly evoking the squalor and exhaustion of laundresses' lives and making the case for the necessity of regulation.[27]

The 1895 bill passed, although nobody but the conservatives celebrated the results. The law was poorly worded and full of loopholes. One commentator acidly remarked that in "other trades, the practice of Parliament has been to take the standard of the good employers, and force the bad ones up to it. With regard to laundries the members took the standard of the bad employers, with the result that the good ones stand in serious danger of being forced down to it."[28] Despite the prescription of a normative ten-hour day, a clever employer could still keep employees for as long as sixteen hours and extract long periods of overtime during busy seasons. Even conventional morality was subverted to the needs of capital and the convenience of the middle class. Lawmakers failed to mandate the Sunday holiday common to other regulated industries. They also exempted small laundries (those employing two or more family members) and religious houses (as a concession to the Irish vote).[29] The strongest part of the act pertained to conditions. It obliged employers to install guards on machines and repair their ramshackle premises. It also empowered the Factory Inspectorate to enforce those rules.

The Liberal Party ascendancy was short-lived. Even before the amendments had been put in place, voters elected the first in a series of Conservative governments that would dominate British politics until the Liberal resurgence of 1906. Attuned to the laundresses' plight, the WTUL and their allies nevertheless continued to fruitlessly push for stronger legislation.[30]

During the intervening years, forces of female reform made their arguments in pamphlets, in articles for journals of opinion, to sympathetic journalists, and to politicians. The annual reports of female factory inspectors occupy the center of this discourse. Far from disinterested bureaucratic summaries, these reports openly advocated further legislative reform and criticized existing regulation. Like many of the documents produced by American progressive reformers, they used evidence gathered in investigations, telling anecdotes, and purported quotes from workers and employers to make their case. Circulated as "blue books," these reports were quoted, paraphrased, cited, and ruthlessly robbed for examples by both American and British reformers.[31]

Because most of the female inspectors in this period also had close ties to women reformers and reform organizations, information also undoubtedly circulated informally. For instance, before her marriage, May Abraham shared a flat with Gertrude Tuckwell, Dilke's niece and Secretary of the WTUL. Lucy Deane, who wrote a widely quoted section on laundries for the Lady Inspector's Annual Report, remained a loyal, dues-paying member of the WTUL after she became an inspector.[32]

As pioneers looking for a foothold in a department dominated by men, Abraham and other "lady inspectors" also had their careers to consider. The newly regulated laundry industry offered an opportunity to carve out an area of expertise different from male inspectors' fiefdoms of textile factories and steel mills. Because the hours provisions of the amendment were so weak and difficult to enforce, safety and conditions inside laundries became the main concern of inspectors. The necessary implementation of machinery turned out to be a professional boon to these women. In the Department of Factory Inspection, "engineering knowledge" of industrial machinery was an important marker of professional status. Women could not compete with experienced male inspectors' expertise on mules and looms and machine tools. But because of the newness of laundry legislation, none of the men knew much about laundry machinery. The women quickly learned to speak authoritatively about mangles and extractors and washing machines.[33]

It took the WTUL a while to figure out how to use this technical expertise. In 1902, the League created a Worker's Legal Advice Committee in conjunction with the London Trades Council. A lawyer hired by the committee represented workers in a variety of disputes with employers. As it turned out, the vast majority of workmen's compensation ("employer's liability" in Britain) cases involved laundry workers.[34] It is likely that these women came to the WTUL's attention because they had a standing arrangement with the lady inspectors in order to screen women workers' complaints against employers and put forward those that seemed most likely for successful prosecution.[35] Because of the aforementioned weakness of the amendment's hours provision, most prosecutable cases involved safety violations. Likewise, until later reforms, success in workmen's compensation cases required proving at least partial responsibility on the part of an employer.[36]

Providing lawyers for workmen's compensation cases and writing articles about the need for laundry legislation were stopgap measures during a period in which the beleaguered Liberal Party was unwilling to waste political capital on the minor business of women workers. In 1906, the tides turned with the reelection of a liberal majority now beholden to the labor movement. A year later, Parliament finally passed a Laundry Amendment that all parties agreed was both fair and thorough. Within the next few years, lawmakers and reformers had put wages, previously sacrosanct, as well as hours, on the table as appropriate for state regulation in the so-called sweated industries, which, at least in some people's minds, included laundries.[37]

Whatever Happened to the Unions?

The Amalgamated Society and the 1891 Hyde Park demonstration marked a high point of organized activity among British laundry workers. In 1893, Clara Collet told the Royal Commission on Labor that she doubted if even sixty members survived out of the original three thousand.[38] By 1895, the laundresses' union had disappeared completely. Later WTUL propaganda claimed acute disappointment over Parliament's failure to pass the 1891 amendments caused workers to abandon hope. The laundry unions more likely succumbed to the same fate as similar unions created in the late 1880s–done in by economic depression and the Trades Union Congress's own conservative retrenchment.[39] If a few laundry workers organized in the following years, their numbers remained very small. The Board of Trade's yearly report on Trade Unions did not even use *laundry* as a category until the second decade of the twentieth century, though the report included unions of female domestic servants, hairdressers, and button hole makers.[40]

In recognition of the problems organizing women, the WTUL created the National Federation of Women Workers (NFWW) in 1906.[41] Under its auspices, laundry unions began to reappear around 1911. Unions also grew because the British Government began using them to organize the distribution of welfare benefits and to set minimum wage rates. In 1913, an amendment to the Health Insurance Act required all workers earning less than 160 pounds a year to join a registered union to receive benefits, spurring even more laundry workers to join the NFWW.[42] World War I increased participation in laundry unions, as labor shortages gave women who remained in the trade more leverage with their employers.[43]

In the prewar period, those few hopeful years between 1889 and 1891 stood out as the high point of organizational activity among British laundry workers. Given the absence of evidence, one can only speculate about the reasons for the lack of activity immediately following that period. In general, women were extremely difficult to organize (with the exception, it seemed, of women textile workers). In 1898, there were approximately 1.5 million women wage workers in Britain. Of these, 116,016–or about 8 percent–belonged to unions. Organizing women textile workers, meanwhile, was the labor movement's great success story. Only 9,546, or .6 percent, of all women workers belonged to non-textile unions.[44]

A comparative perspective only makes this situation more puzzling. American workers were far more active than their British counterparts, despite greater hostility from employers and the courts and a notoriously unsupportive male-dominated labor movement. This discrepancy was possibly because British women workers saw legislation as a better solution to their problems than ineffectual unions.[45] The Factory Acts pre-dated most American protective legislation and provided more reliable coverage than most state laws passed during the progressive period. However, the weakness in the hours provisions of the act meant that women workers in some progressive states like Illinois, California, and Massachusetts had shorter legally mandated work weeks than their British counterparts (although the laws were applied more sporadically because of court challenges). On the other hand, until the formation in Britain of the Laundry Trade Boards, after World War I, British legislators did not address the question of minimum wages for laundry workers—an issue of primary concern to workers and a matter that unions could do something about. The WTUL and other reform groups believed that organization and regulation should go hand in hand, mutually reinforcing each other.[46]

A more provocative explanation for these national differences lies in the differing composition of British and American unions. The Amalgamated Society of the early 1890s was a single-sex union. This strategy protected women from male workers who might use the union to their own ends. Because of the nature of women's work lives, single-sex unions were often ephemeral. Mixed American unions during 1903–1914 were far more stable, but women members were still continually betrayed at the bargaining table by male union members. In Britain, the NFWW put forward the general union composed of women workers as a solution. Americans rejected this model because it was generally unacceptable to the AFL. In short, unionization was theoretically the best solution to women workers' problems, but in practice neither the British nor the American solution really worked for laundry workers. Who could blame women who worked sixty-hour weeks, returning home for a second shift of childcare and housekeeping, for not wanting to invest their Sundays and evenings in a cause from which they could hope to gain little?

The American Case

American efforts at reforming the industry, like those in Britain, placed an emphasis on intertwining legislation and organization. However, these similarities disguised profound national differences born of context and chronology. AFL-linked unions began later than similar unions, in a political climate in which many employers and trade associations bitterly (and sometimes violently) opposed both unionization and regulation. Legislative efforts had to be carried out on a state-by-state basis, and they had to be defended against constitutional challenges. As in Britain, the American WTUL (modeled on the British WTUL) would come to play a central role in organizing laundry workers, particularly through its strong New York and Chicago chapters. However, the organization was not founded until 1903, making it a relative latecomer to the business of organizing laundry workers.

In the late 1880s, efforts by the Knights of Labor to organize women and unskilled workers paralleled Britain's New Unionism. With the exception of the special case of Troy collar laundresses (women who finished the stiff collars newly manufactured in the factories of Troy, New York), these movements resulted in only a few short-lived locals.[47] The combination of the Knights' collapse, conservative gender politics in the emerging AFL, employer backlash, and economic hard times that began in 1893 all contributed to the apparent disappearance of laundry workers' unions during the 1890s. Concentrated efforts such as the dramatic organization attempts of Atlanta-area laundresses seem to have had little effect on either domestic or commercial laundry workers of the next decades.[48]

The power of the AFL in the American labor movement dominated the history of laundry unions in the early twentieth century. Middle-class reformers felt compelled to cooperate with the federation's policies, endorsing mixed-sex unions that often gave priority to the needs of male workers. Reformers hesitated to question these policies, for fear of giving the impression of breaking ranks and jeopardizing the resources the national federation offered. Between the demise of the Knights of Labor, in the 1880s, and the rise of the Congress of Industrial Organizations (CIO) unions, in the 1930s, no alternative to the AFL-sanctioned locals existed.

The Shirt, Waist, and Laundryworkers' International Union was the AFL national. Organized in 1898 in Troy, New York, as a local union and chartered in 1900 as a national union, it was a conservative, male-dominated organization with its roots and strategies oriented towards workers who laundered newly manufactured collars.[49] Surviving materials from its earliest years laid out a platform dedicated to: organizing a national trade union; securing a "universally equal and just rate of prices without strike"; and maintaining a death benefit for all members. That last clause made the union particularly unappealing to women who could not afford the assessment and did not think it a wise investment because they most likely would not leave the laundry, after their relatively short work lives, in a box.[50]

The Laundryworkers' International Union broadened its membership base through the activities of AFL organizers, but it continued to cling to its identity as first and foremost a union for male workers in manufacturing laundries. The limitation of jurisdictions seems partly to have been a strategy for avoiding union involvement in situations offering little possibility of success.

The presence of male union members, the support of the AFL and other organizations, and the confrontational nature of the American labor movement all show that American laundry unions exhibited more inclination than their British counterparts to use high-profile, citywide strikes to win concessions from their employers. From 1903 to 1912, thousands of workers walked away from their places of employment, in a series of large and small strikes, leaving linen rotting in washing machines and inspiring desperate customers to break into laundries in an attempt to retrieve their bundles. Or, as a *New York Times* editorial cartoonist imagined it, inspiring resourceful bachelors to take a brush and whitewash to their collars and cuffs.[51]

Laundry Strikes

The first large-scale uprising in the laundry industry took place in Chicago. As the city with the second largest number of laundries in the country, with a concentration of large factory-type laundries that serviced the hotel and travel industry, and an active labor movement, Chicago was primed for action.

On the morning of April 23, 1903, seventy-five women and girls from the company laundry in the community of Pullman, on the outskirts of Chicago, responded to a prearranged signal and walked out of the laundry. Their gesture was probably not entirely a surprise to the management of the huge sleeping-car company that controlled the company town. The laundry, set up in 1892 as a means of providing em-

Fig. 17. A *New York Times* cartoonist imagined the response of desperate men deprived of laundry services during the 1912 citywide strike. *New York Times* (7 January 1912).

ployment for the female relatives of car-works employees, had not been immune to the labor unrest that had been rocking the town for the last decade. Pullman laundresses had also walked out in 1894 as members of the American Railway Union.[52]

This time they were striking, said the *Chicago Tribune*, for a "nine hour day with ten hours pay." Four days later, the same paper reported that "the girls that struck at Pullman Palace Car Laundry will visit the Chicago laundries today in an attempt to discover which of them is furnishing towels and linen to the company. Should they be successful, they will ask the employees of the firm to join the strike."[53] The effort was already too late for the Pullman women, for the company had set to work converting the laundry into a machine shop and arranging to have future laundering done in St. Louis, Omaha, and other cities.[54]

Even though they had failed to come to terms with the giant sleeping-car company, something in these women's courageous gesture struck a chord in other workers. Threats of organization that had been reaching the ears of Chicago laundry owners for months now coalesced.[55] The papers reported that more than two thousand laundry workers had descended on union headquarters, at 122 LaSalle Street, to reject an offer made by their employers for a wage increase and improvements in work conditions.[56] Employers also gathered and arranged a lockout. By May 4, 146 of the 170 large steam laundries in the city had shut down, whether or not their workers were on strike.[57]

As the newspapers detailed the day-to-day progress of the strike, their accounts reveal the divisions and politics of the Chicago labor movement in this period. The exact dynamics of the union are unclear, but the AFL apparently had a hand in organizing workers for several months before the strike.[58] Despite these connections, the Chicago Federation of Labor (CFL) failed to recognize the strike until it had been in progress for almost three weeks. One reason for this is that the laundry workers acted without the sanction of the CFL, which had recently passed a resolution requesting unions to confer with the central executive board before going on strike.[59] Representatives of the union did not attend the May 3 meeting of the Federation and no mention was made of the union or the strike. The late entry of the CFL coincided with the involvement of the Teamsters and the proposal that the Chicago Board of Arbitration, the "adjustment body of the Teamster's Union," attempt to settle the strike.[60]

The evidence and the eventual outcome suggest that the laundry workers and the CFL disagreed about means and ends, and therefore the CFL bided its time, knowing that without substantial strike funds or the support of other workers, the laundry workers would have to come to terms. Ann Metzger, described in the papers as the secretary of the union, displayed more confidence. She declared to reporters that the laundry owners were weakening: "It [the lockout] won't last long. Every one of the bosses will be praying for a chance to sign our little agreement."[61]

Meanwhile, the city began to feel the effects. Restaurants made do with paper napkins and hotels worked to convince their guests to sleep an extra night on dirty sheets. The *Chicago Record-Herald* reported that Chairman Barkey of the executive committee of the Chicago Laundrymen's Association appeared at the May 8 meeting wearing a sweater over his wrinkled shirt. Several of his fellow members were spotted wearing puff ties to hide their soiled shirt fronts.[62] Laundrymen punned that "the laundry strike is all the more serious from the fact that Chicago has no flatiron building."[63] The *Chicago Tribune* headlined a May 6 article on the strike: "All but the Public Enjoying the Strike: Laundry Girls Pass Day Singing and Dancing." "Settle and return to work?" queried one of the strikers, "I guess not. We are having the time of our lives."[64]

Laundry workers had also managed to cut off aid from the outside. Work sent out of town was returned unwashed, with the word that laundry workers in Milwaukee, Joliet, Racine, and Elgin had refused to accept it.[65] The *National Laundry Journal* reported that "the Chinese Laundries were doing a land office business for a few days, but the strikers put a stop to their sudden prosperity. A strong, husky chap stands before every Chinese laundry and looks at 'John' when 'the man with the bundle' enters." The journal claimed the counterman was so intimidated as to refuse the bundle, declaring "no washee doing stwike."[66] John Alexander Dowie, the virulently antiunion leader of a religious community just outside Chicago, fared better. "Zion City is protected with a large and well equipped laundry and since the Chicago linen famine commenced, it has been running night and day."[67]

The strike lasted throughout the month. With funds dwindling, the leadership of the local capitulated control to the CFL, the international union, and the Teamsters. The editor of the *National Laundry Journal* later mused about this development. He noted that the international had no treasury fund: "This being the case, we cannot see why the laundry workers are so anxious to connect themselves with the International body."[68]

In the end, Mr. T. E. Wilson, president of the local laundry worker's union, negotiated a settlement. Closeted with the president of the Chicago Laundrymen's Association, Pliny F. Munger, and members of the Illinois Manufacturers Association, Wilson worked out an arrangement that made the female laundry workers so angry that they threatened to go back on strike.[69] In addition to resenting Wilson's inability to negotiate any substantial improvement in wages, hours, or conditions for the female employees, the women were also angry that the agreement did not include recognition of the union, leaving them in no better bargaining position than they had been at the beginning of the strike. Male employees fared a little better in terms of wage scales, but the real beneficiaries were the laundry wagon drivers, who had little role in the initial strike. The drivers had the Teamsters Local 22 hammer out a separate agreement that held up the settlement of the strike for an extra week.[70]

After a month of panic, the laundry owners felt smug when the strike came to an end. They attributed their success to the firmness of their resolve. The editor of the *National Laundry Journal* sounded a note of pity for the female laundry workers, who had composed a large proportion of the strike force. He suggested that they were understandably naive and had been misled and misused by power-hungry unionists. (Better they should place their faith in the good intentions of their employers!)[71]

San Francisco was the labor movement's great success story of pre–World War I laundry organization. Union officers pointed with pride to the process through which San Francisco steam laundries had been transformed from some of the most exploitative in the country to closed shops with relatively high wages. As in London, the origins of the San Francisco union were tangled up with the push for protective legislation. In the summer of 1900, the California State Labor Commissioner and several newspapers began to receive letters demanding an investigation of conditions in San Francisco laundries. Many of the letters were anonymous and seemed to be from workers who were afraid of losing their jobs.

In response to these letters, the *San Francisco Examiner* sent a woman reporter to work undercover in one of the city's laundries.[72] Her investigations revealed that even in comparison with steam laundry workers elsewhere in the country, San Francisco workers toiled under terrible conditions. Laundry owners had instituted a system of dormitories for female workers, with squalid rooms and poor food. After deductions for room and board, this system allowed them to pay workers a miserably small wage of eight to ten dollars a month. Women living at home were earning scarcely more: about $17.50 per month.[73] In response to these findings, the County Board of Supervisors limited nightwork and restricted hours to thirteen a day (sixteen had been the previous limit), on the grounds that such long hours were detrimental to the health of workers.

All of this inspired the male laundry workers to apply to the international union to form a new local. They intended for its membership to be limited to men, but (to its credit) the international refused to grant the membership unless women were included. A woman organizer was hired to solicit members for several months, until it was decided that enough workers had been signed up. In 1901, the laundry owners acceded to the union's demands, abolishing the boarding system and raising wages.[74]

Through the next few years the union worked quietly, organizing workers, building a strike fund, and gaining worker confidence by providing relief during the 1906 earthquake. In April 1907, the union members took advantage of the disorganized state of the local laundry owners to strike again. Their demands included an eight-hour day and a substantial increase in wages. When the demands were refused, eleven hundred workers from fourteen laundries walked out.

It was a strategy worthy of the LNA. The *National Laundry Journal* ran an incensed article in which the author questioned the civic-mindedness and team spirit of the strikers for trying to take advantage of the misfortunes of their employers. The article also made the usual suggestion that union members had been misled by their leaders.[75] The strike lasted eleven weeks. The eventual agreement negotiated a gradual decrease in hours over a three-year period to an eight-hour day. Other concessions included a fixed scale of minimum wages, safety devices for all machinery to protect workers from injury, and the negotiation of closed-shop agreements.

The union contract was honored and the union persisted. However, that simple success story disguised two dirty secrets: First, employers tolerated the union because union leaders had agreed to help them drive Asian and French competitors out of business through harassment and intimidation. Racism created a bond that overrode the divisive effects of gender and class antagonism.[76] Second, as in Chicago, male union members had negotiated an agreement that benefitted male workers far more than their female counterparts. In the union wage scale, the highest-paid workers were not the ironers but the markers and washers. In laundries in other parts of the country, male washers were seldom on the top of the pay scale and most markers were women. But in San Francisco, after the strike, male workers managed to maintain a monopoly on marking jobs–in part because the union agreement stated specifically that "women markers and distributors must be paid the rate of wages paid for men employed at the same work."[77]

When Lillian Mathews researched the San Francisco union around 1912, union leaders explained the rationalization for their wage scale. Washers were skilled workers because the work requires "skill and good judgement," to avoid overfilling or underfilling the washing machines with water. Within the union, Mathews found "internal friction is not uncommon between the men and the women over the conduct of affairs. . . . As it is now, the men are inclined to assume a domineering attitude, which makes it unpleasant for the women to speak freely, and discourages those girls who are naturally timid from asserting their opinions."[78]

In an open-shop town, the women would have left the union. The closed-shop agreement, however, obliged them to belong in order to keep their jobs. Mathews tendered an explanation as to why the greater number of women had not taken control of the union: "In spite of the fact that the women in the union outnumber the men almost five to one, men are always elected as delegates to the State Federation of Labor." She thought this situation was "due to the fact that the women will not stand together on a woman candidate, those in one laundry being unwilling to vote an honor to a woman working in some other laundry, each laundry group forming a clique of its own. Rather than see an honor go to another clique, the women will vote for the man."[79]

The San Francisco case was unusual. In general, women workers would turn out for a specific cause (strikes, for example), but they did not stay in the union. In 1908, the U.S. Department of Labor did a special census of women in laundry unions. Only two hundred female union members were identified in the entire state of Illinois–or about 2 percent of the total number of workers in Illinois steam laundries identified by the census of manufacturers the next year. New York had a similarly dismal 3 percent, and even California, fresh from the triumph of the San Francisco strike, had only 21 percent.[80]

Until 1912, New York, home to the largest number of laundries in the country, had been spared any large-scale action by laundry workers. But on January 1, 1912, New York laundry workers began their first citywide strike. The nature of the strike and the strikers reveals the changes that had overtaken the industry and the nature of organization in the trade over the previous ten years. This time, the strike was precipitated by the walkout of 650 workers from Local 126 of the United Laundry Workers, affecting as many as twenty-five large plants. These were not skilled workers but operatives who did rough dry work subcontracted by smaller laundries. Despite the fact that nearly 450 of the workers were women, the strikers' demands reflected the priorities of the male-dominated union. Topping the list of demanded salaries was an astonishing twenty-five dollars a week for the washmen who handled the collar and cuff machines. In contrast, mangle girls were to receive a minimum of six dollars a week–less than many were already receiving in many New York laundries.[81]

The *New York Times* reported that the strike was in response to negotiations to form a trust to control businesses that did rough dry work in the Bronx, Brooklyn, Manhattan, and part of New Jersey by buying up companies, issuing stock, and eventually replacing management. Union leaders felt, probably rightly, that once the trust was in place, it would be impossible to wring any concessions or even recognition of the union from such a powerful concern. In speaking to the reporter for the *Times*, union spokesman William Armour compared conditions in New York to those in San Francisco, suggesting that the strikers were taking the success of the San Francisco union as their inspiration.[82]

What had been a clearly drawn situation involving two opposing, well-defined parties quickly began to degenerate into a more complex and ambiguous situation. By the second day of the strike, the ranks of strikebreakers had swelled to two thousand, while other workers rushed to join the union. Other locals went out in sympathy, some of which, such as Local 34 on the East Side, counted among their members not only employees but also some employers in smaller laundries.

A committee of hand laundrymen also met with strikers and agreed to close sixteen hundred hand laundries to drive the rough dry owners "into submission."[83]

Though the strike had begun as a response to the strategies of the male-dominated union, those interested in organizing women began planning a strategy to reach their constituency. In particular, the WTUL (which had a frustrating history trying to organize women laundry workers in New York) saw this as an opportunity to reach those previously unreachable.[84] Much had changed since the league had disbanded its laundry committee in 1906, citing a lack of progress in organizing. Successful involvement in a series of shirtwaist makers' strikes had buoyed league members' confidence that New York women could be successfully organized. Moreover, the laundry industry had been the object of much public scrutiny resulting from exposés in the press, suggesting that there would be public sympathy for the strikers.

On January 6, the NYWTUL began organizing women workers who were not part of the union. They held a "parade for girl strikers," which culminated with speeches by Mary Dreier, president of the WTUL of New York, and the prominent British feminist Sylvia Pankhurst.[85] Soon thereafter, the forces of middle-class feminist reform went to work to publicize the plight of these women workers. The WTUL declared that supporting women laundry workers was a moral issue and that ministers should be doing more to promote their cause from the pulpit. Mrs. Helen Marot, secretary of the NYWTUL, brought two workers to the Episcopal Church of the Ascension to describe their plight. A fifteen-year-old girl told the State Board of Arbitration about the "slavish" conditions in the laundries. She walked thirty-two blocks to work, and one of her fingers had been crushed in a mangle accident.[86] Times had changed. Women workers may have had less sway over their employers, but they now had more pity on their side. Reformers used public opinion to ride on the coattails of male unionists.

The outcome of the New York strike was disappointing for all the workers involved, however. When it ended on January 31, only six laundries had agreed to recognize the union and to pay higher wages—the rest had hired scab labor to replace the strikers.[87] The crowd of women workers that had signed with the unions and flooded the streets to hear Pankhurst and Dreier speak now disappeared, going back to their mangles and shirt presses.

The New York strike stands apart from other British and American strikes of the period because it took place in a state that prided itself on progressive labor legislation. Many members of the public believed legislators had already fixed the conditions that striking laundry workers decried. In the aftermath, there was public outcry for an investigation into the strike's causes. Four days of public hearings added to the conclusion that too many laundry workers endured horrifying work conditions and low wages. However, in the end the committee did little to ameliorate conditions in the laundries.[88]

Muller v. Oregon

The point of a public hearing was not just to shame laundry owners into accepting unions or treating their workers more humanely. Progressives hoped to use this spectacle to ensure existing laws were enforced and to impose new regulation on the industry. Heartened by the Supreme Court's 1908 decision in *Muller v. Oregon,* they believed it was now more possible to create effective and permanent protective legislation for women (if not yet for men).

Over the course of the twentieth century, *Muller v. Oregon* has become one of the most famous cases in constitutional law. In upholding the right of states to regulate women's work hours, the majority opinion enshrined class-based legislation–in this case, ruling that women's real or potential role as mothers overrode their constitutional right to decide as individuals how long and under what conditions they would work. As such, *Muller* still provides fodder for debates about protective legislation and sexual distinctions in the law. The case is also remembered for pioneering the use of "sociological" arguments to support social legislation. Only a small portion of the so-called Brandeis Brief cites legal precedents. Instead, it extensively quotes various experts on women's industrial labor, describing the conditions under which women worked and the social and biological effects of that labor.[89]

Muller is also a case about a laundry. As such, it provides further insight into the special role laundries and laundry workers played in efforts to reform and regulate the conditions of women's industrial labor.

From a strictly legal perspective, a test case could have been based on violations in any number of other industries. However, by 1908, the Supreme Court had made it clear that it would not uphold the constitutionality of protective legislation for male workers. In the 1905 case *Lochner v. New York,* the Court declared that if it reified a special law for male bakers, other men in dangerous or unhealthy occupations would also have to be protected–a step the Court was unwilling to take.[90] In the years that followed, protective legislation became nearly synonymous with laws specific to women workers. No one needed to explain that a case about a laundry was a case about women. Moreover, nearly a generation of investigation into women's work (laundry work in particular) provided a huge body of literature articulating the physical and social effects of such labor in terms of gender. The "Brandeis Brief" was built quite literally on the work of the British lady inspectors and their Continental European counterparts.

Bringing the case *Muller v. Oregon* to the supreme court was a collective effort, but Florence Kelley acted as its prime mover. Former Chief Factory Inspector for the state of Illinois and general secretary of the National Consumer's League since 1899, Kelley translated her own frustration as an inspector and as an advocate of women workers into a campaign to create stronger protective laws.[91] Kelley sought out an opportunity to challenge the precedents that had undone previous efforts to create and maintain legislation. In 1907, she found one.

Sometime that year, Curt Muller, a laundry owner in Portland, Oregon, had received an unwelcome visit from a factory inspector that resulted in a summons to appear in court. It seemed that by working his female employees more than ten hours a day, he was violating a statute of the state labor code. Inspectors themselves were, as a rule, overworked, and so Muller had probably not expected that he would ever be singled out for enforcement of a law that had been on the books for several years. He protested as much to Judge Alfred Sears of the Circuit Court of Multnomah County. His lawyer cited the precedent of *Lochner v. New York,* drawing from that decision the arguments that Oregon law denied Mr. Muller equal protection under the law, that it exceeded state police powers, and that it abridged his privileges and immunities as a United States citizen.[92]

Judge Sears ignored Muller's appeal to precedent and offered him the choice of a ten-dollar fine or five days confinement in the county jail. He concluded that *Lochner* did not apply because it was a case involving male bakers, while the Oregon law was designed specifically to protect the health of female workers. Though little is known about Muller, stubbornness must have been one of his character traits. He refused to hand over the nominal ten-dollar fine and be done with the matter. Instead, he appealed the case to the Oregon Supreme Court–with the same results. Still unwilling to back down, Muller then applied to the United States Supreme Court for a hearing.[93]

At this point, Muller's sense of injustice became incidental to the proceedings, as two powerful groups latched on to the case for their own ends. On one side, the Oregon Manufacturer's Association provided backing for Muller. On the other side, the forces of middle-class reform marshaled themselves. The Oregon Consumer's League (the members of which Kelley had cultivated during a 1906 trip through the Northwest) notified the national headquarters that *Muller* would be an excellent test case.[94]

In retrospect, the choice to focus on laundry workers in the battle over protective legislation for women seems obvious. However, Kelley soon discovered what British reformers already knew–that the young, tractable laundry operative in need of help had her diametrical counterpart: the Amazonian washerwoman who, in the minds of some, could easily help herself. In her biography of Kelley, Josephine Goldmark (who played a major role in constructing the brief and was also Brandeis's sister-in-law) told a story of Kelley's unsuccessful first attempt to secure a lawyer to argue the case. According to Goldmark's account, the male board members of the Consumer's League had scheduled an appointment with Joseph Choate, a leading member of the New York Bar. Choate was puzzled by the whole context of the case. Goldmark described his response: "'A law *prohibiting* more than ten hours a day in laundry work,' he boomed, 'Big, strong, laundry women. Why shouldn't they work longer?'"[95] There is only Goldmark's word that this is what Choate actually said, written down forty years after the fact, but the sentiment is symbolic of a whole older order that did not realize or appreciate what Goldmark and Louis Brandeis, the lawyer finally chosen to argue the case, were about to do. In more than one hundred pages of the "Brandeis Brief," they transformed the cultural image of the laundress from a big, strong Irishwoman, the traditional Amazon, to a helpless, weak, ignorant mother-to-be.

The brief was researched and written by a team of women from the Consumer's League led by Josephine Goldmark. Its composition illustrates the transatlantic character of reform during this period. At the Columbia University Library, Goldmark went directly to the Parliamentary Papers. "It was there in the successive reports of British Factory Inspectors and British medical commissions, beginning with the First Children's Commission of 1833, that we found what we were seeking," she later wrote.[96] Like a long string of British experts, Goldmark and her team would quote Lucy Deane's opinion on mechanization in the industry and the inspectors' carefully gathered facts about accidents and long hours.[97]

In order to successfully argue the case, Louis Brandeis had to communicate explicit limits of the protection to be extended. Though the enduring crux of his argument was centered on the biological differences between the sexes, there was also a careful attempt made to describe nonphysical differences between genders: Brandeis warned that long hours impaired women's moral judgment. Overworked mothers neglected their children and their duties in maintaining the home.[98]

The debates over protective legislation, over the Factory Acts, and even within unions all took place among the more privileged, effectively leaving those most concerned out of the debate.[99] In some ways, *Muller* is an extreme example, because it is couched not only in the language of the law (Kelley had to hire a specialist just to enter a debate on a subject in which she had immense experience), but also, through Brandeis's stroke of genius, in the language of "science."

Brandeis made an appeal to the court that, like the brief, contrasted what Goldmark described as "Hyperion to a Satyr": good with evil (and incidently, helpless women with big, strong laundresses).[100] In his opinion for the majority, Justice Brewster responded with what would become the classic justification for protective legislation for women:

> Even though all restrictions on political, personal, and contractual rights were taken away, and she stood, so far as statutes are concerned, upon an

> absolutely equal plane with him, it would still be true that she is so constituted that she will rest upon and look to him for protection: that her physical structure and a proper discharge of her maternal functions—having in view not only her own health, but the well-being of the race—justify legislation to protect her from the greed as well as the passion of man.[101]

The significance of *Muller* became apparent to workers, reformers, and laundry owners only gradually. In the immediate aftermath, it was understood primarily as having reified laws already on the books in a handful of states. The editor of the *National Laundry Journal* gave the decision only a scant paragraph, warning laundry owners that they should avoid challenging existing laws and instead put their energies into other means of staying ahead of the competition.[102] Workers and reformers celebrated quietly and went about the business of lobbying state legislatures to pass new laws and enforce current ones.

Lulu Holley Joins the Union

A strange group of lobbyists descended on the Illinois State House in June 1909 to press for the passage of an eight-hour bill for women workers: Elizabeth Maloney and Anna Willard, secretary and president, respectively, of the Waitresses' Union, Local 484 (Chicago), Agnes Nestor, international secretary of the Glove Workers' Union, and Lulu Holley, identified in later accounts only as a "laundry worker" and "member of the laundry worker's union." The waitresses were there because their union had prepared the bill in cooperation with the Chicago WTUL and a very young Harold Ickes.[103] Nestor was one of the most prominent female working-class labor leaders in Chicago. Lulu Holley was one of the four because in the wake of the Supreme Court decision on the *Muller* case, laundry workers had become symbolic of the need for state regulation of women's work.

The previous ten-hour law in Illinois had been ruled unconstitutional by the state's supreme court in 1895, during the period Florence Kelley had been Chief Factory Inspector for the state. Its failure had helped motivate her to pursue *Muller*.[104] Meanwhile, members of the Chicago WTUL and other reformers had been carefully planning to introduce a bill restricting women's work hours into the Illinois legislature. To aid in the passage of this bill, Brandeis and Goldmark had prepared a new brief modeled on *Muller*. But they also decided to ask the legislature to restrict workers' hours to ten a day rather than the eight that working-class members of the WTUL wanted. Lulu and company had initially gone to Springfield alone, hoping to push an eight-hour bill through, but they had found it difficult to access lawmakers. In the end, middle-class reformers reentered the picture, eventually obtaining passage of the more conservative ten-hour law.[105]

Several years later, Holley's compatriot Elizabeth Maloney wrote a brief note for the Chicago WTUL's biennial report: "The Department of Factory Inspection was granted an increased appropriation by the last General Assembly for five additional inspectors." The WTUL requested that "at least one of these additional inspectors appointed be a woman, and recommended for the position Miss Lulu Holley." Lulu took the exam for general inspector, placing first among her group of examinees. She thereby entered a job Florence Kelley had helped establish, joining the profession of Lucy Deane and May Abraham.[106]

Like most other women laundry workers, Lulu is a shadow figure in the historical record, leaving behind no explanation of her actions. She must have drawn some conclusions from her experience with the Chicago WTUL, however, recognizing that it is good to be helped by reformers, but it is better to *be* one of them, obliged only to visit the workplaces of the women one is trying to help.

✕

In both Britain and the United States, efforts to improve working conditions for women laundry workers focused on organization and regulation, illustrating a transatlantic culture of reform. In the United States, those efforts built on the movements and institutions created by British reformers. During this period, however, neither attempts at unionization nor those for state-mandated reform could be considered a glittering success. Laws for factory legislation did improve the comfort and safety of work conditions for some workers, particularly in Britain. Until 1907, though, rules limiting hours were so ambiguous as to have little effect on many British employers. In America, laws in some states were more rigorous but had to withstand constitutional challenge and required effective enforcement. Union activity waxed and waned. Ironically, American unions seemed to have prospered more than their British counterparts, despite the greater resistance to unionization on the part of American employers.

It is difficult to assess the impact of regulation and unionization on the industry itself. For political reasons, reformers liked to claim responsibility for any positive changes in the industry while simultaneously crying for stricter regulation. If laundry owners talked about the effects of regulation and unionization at all, their conclusions were mixed. In the prewar period, it was difficult to disaggregate the economic and regulatory incentives for rapid mechanization and growth in scale. In Britain by the 1920s and the United States by the 1930s, that ambiguity disappeared as stronger laws and stronger unions undeniably changed the way individual businesses were run and altered the demographics of the industry as a whole.

Perhaps more significantly, the workings of the laundry industry were no longer a mystery to the public. Reformers and muckrakers made sure that the worst labor practices and most awful working conditions were detailed in newspapers and magazines. Demand, however, seemed as steady as ever, and so, with the steady tenacity that they liked to think characterized their breed, laundry owners set about to control the damage, confident that the next decades would bring ever-increasing business.

PART II

Confronting Maturity, 1920–1940

Production and Consumption during the Interwar Period

By the 1920s, power laundries had long ceased to be a new industry. In some successful businesses, a second generation presided over their fathers' laundries.[1] Men who had entered the trade as hydro-boys, washmen, or drivers now spent their last years in the foreman's chair. Even individual women workers who had never been expected to stay long grew too old to operate the machines they had first encountered as girls. The institutions of the laundry community also became increasingly entrenched. Annual dinners and conventions came and went. Stacks of trade journals grew tall, gathering dust in the corners of offices. Middle-class housewives and single men continued to send their laundry out, as they had seen their mothers and fathers do.

Maturation brought with it both a sense of permanence and a growing awareness of the industry's technical and organizational limits. Technologically, this period marked an end to rapid mechanization. All the major processes that could be mechanized had been mechanized. More and more, new machinery offered only incremental improvements. This technological stasis left laundry owners with few options in solving the problem of rising labor costs. Minimum wage laws and union demands drove up wages in Britain and in some progressive states like New York and California. American laundry owners found themselves paying workers more and looking for new sources of employees as stricter immigration laws dried up the pool of immigrant labor. In Britain, the implementation of minimum wages through the formation of a laundry industry trade board, in 1918, effectively doubled the amount many workers were paid. Almost everywhere, protective legislation and changing work patterns gradually put an end to the sixteen-hour days and all-night stints that had once characterized the industry.

Even those laundry owners cloistered in small towns and stable businesses could not help but notice that the world around them had changed since prewar days. Some could see that transformation in the composition of the bundles unpacked by markers. The separate, starched white collar, essential element of respectable male attire for almost a century, was disappearing. The soft-collared shirt that replaced it could more easily be washed and ironed at home.[2] The size of bundles also shrank as customers increasingly looked for ways to economize. Some sent only part of their washing. Others employed more complicated strategies. One British patron sent the same number of pieces but in smaller sizes: The laundry charged less for the narrow linen runners she substituted for tablecloths, and small guest towels were cheaper to launder than the large ones she had always used. She told an interviewer she had also taken to wearing plain lingerie that did not require ironing and could be washed at home.[3] Some longtime customers who had ceased sending their laundry out during the war simply did not start again.

A telephone survey conducted by one American laundry revealed shifting patterns of consumption in statistical terms. Of the people who had discontinued laundry service, 35 percent said they had done so after buying newly available electric washing machines. The remainder either could no longer afford laundry service or were dissatisfied with the quality of work and the slowness of the laundry's five-day delivery plan.[4]

Consistent with business ideology of the time, public discourse in the industry most often portrayed social and technological change as harbingers of economic growth. American laundry owners, in particular, insisted on ballyhooing the industry's unlimited future. "A Billion Dollar Business by 1930!" trumpeted an LNA slogan adopted in 1924. Industry representatives liked to claim this represented a realistic goal, since, by their estimation, laundries currently reached only 10 percent of their potential customers. They hoped new services, better marketing, and new cost-cutting techniques would draw in that other 90 percent. They prescribed utilizing "modern business development methods" to more than quadruple American laundry owners' receipts over the coming decade.[5] Persistent economic problems and rising wages made many British laundry owners less optimistic. Like their American counterparts, however, they believed in the power of new techniques to bring steady growth.

The reality was rather more complicated. Innovation and systematic management had clearly helped individual laundries survive and prosper, even during the depression. Nevertheless, none of these strategies could prevent the overall decline of the industry that began to take hold in the 1930s. The industry continued to flourish as long as the only viable alternatives to laundry service were the washerwoman or the washtub. But by the mid-1920s, the spreading availability of electric washing machines threatened not only to limit new customers but to take away old ones. Once consumers had the option of buying a washing machine, they became far less tolerant of lost or damaged items, week-long waits for clean clothes, and other delays caused by unpunctual laundry drivers. The Great Depression worsened this pattern and hastened the transition to washing machines.

Rising labor costs, technical limitations of mechanization, and competition from washing machines created a perplexing problem. To encourage growth and solve some of the inherent difficulties of the industry, individual laundry owners, trade journals, and trade associations explored a number of social and technological strategies. As had always been the case, they looked for answers not only from within their own ranks but also in the practices of other industries. In the 1920s, the techniques of American big business often provided exemplars. "Progressive laundry owners" tried to emulate the strategies of mass production by offering services that eliminated the handling of individual items.[6]

The revolution in mass marketing, with its advertising techniques and efforts to better understand the consumer, also became a source of new sales methods and ideas. In addition to their more traditional functions of sharing information and controlling competition, laundry trade associations became a means of replicating the organizational capabilities of large corporations for research, advertisement, and negotiation with the state. Adoption of these techniques was not just about economic rationality. By emulating the business icons of the age and embracing strategies and attitudes considered modern and progressive by the broader business community, laundry owners helped ease their own sense of inferiority and backwardness.

Progressive laundry owners imagined a future for the industry characterized by highly mechanized plants, standardized procedures, and services and prices that were consistent from laundry to laundry and town to town. They wanted consumers to find opening a bundle back from the laundry as predictable as opening a bag of Wonderbread from the A&P.

Things Go Wrong

In 1924, Georgiana Smurth-Waite, a home demonstration agent for the University of Idaho Extension Bureau, told a convention of Utah and Idaho laundry owners she had discovered that housewives were "crying for laundry services." As director of a four-state survey of female consumers, she had also found that these women were very wary about patronizing their local laundries. Some women avoided laundries entirely because "the cost is high, the loss is excessive, colored clothes are faded and fabrics weakened, and in many cases returned to the owner with holes in them, looking like they had been chewed." Other women continued to send their work out, but they expressed their ambivalence by hesitating to include heirlooms or other irreplaceable items.[7]

Laundry Age reprinted Smurth-Waite's speech for the benefit of a wider audience. It typified a new transatlantic genre of trade journal articles promising laundry owners insight into "what the housewife wants." These articles proposed that continued prosperity in the industry depended on better service, increased "goodwill" on the part of consumers, and savvy marketing of existing services. Because this discussion focused on everything that could go wrong when one sent one's clothes to the laundry, such journal articles also help the historian understand why consumers might be "crying for laundry services" and yet continued to choose other alternatives even if they involved more work.

Although laundries attracted a wide variety of items from different consumers, including that perpetual mainstay, the "bachelor bundle," both British and American rhetoric assumed industry representatives were male and consumers were female. Significantly, most of these articles did not offer specific technological solutions to the problems of faded goods and holes chewed in table cloths–though that kind of advice could be found elsewhere. The central problem this genre claimed to address was a perceived failure of communication between male laundry owners and female customers (most often described as "housewives").

Gender stereotypes heightened the contrast: Female customers were portrayed as ignorant and misinformed, in need of education about the methods and virtues of commercial laundry services. Meanwhile, laundry owners lacked sympathy for the customer's point of view and could too often be found playing pool rather than supervising their female employees.[8] "The modern laundry from the modern woman's point of view is a fearsome place" one article explains. "It consists of a large building containing mysterious machines which cleanse her linen in some magical way–a way which ruins her choicest linens without any hopes of redress. It is up to every laundryman to dispel this ignorance, and to create a better impression amongst customers."[9]

To get the "woman's point of view" (as one LNA column called it) on laundry services, laundrymen asked clubwomen, wives, sisters, and mothers to address conventions and write articles describing their experiences with laundries. Professional women, particularly home economists, were also hired to serve as mediators between female customers and laundrymen.[10] The strategy was predicated partly on the cultural assumption that women participated in daily experiences and conversations that men could not understand or be privy to. Somewhat perversely, laundry owners seemed to believe that laundry fell into this category.

Home economists' professional credentials allowed them not only to explain laundries to women but also to translate complaints that might otherwise be understood as ignorant, feminine nagging into authoritative advice about how laundry owners should conduct their business.[11] Industry leaders recognized that the way many laundry owners handled customer complaints was a major source of ill will. Laundry owners realized that some amount of loss and damage inevitably resulted from laundry procedures, but they arrogantly believed that most complaints were either the result of customer error or an intentional effort to deceive. Others ignored complaints, hoping they would go away, or segregated customers into lists of those whose complaints they would honor and those whom they chose to ignore.[12] One British article suggested that many laundry owners had a distorted view of losses and damages because their own laundry always emerged unscathed. "The 'boss's' mark is known and looked for by all and sundry, and special care is exercised to ensure that every article bearing this mark shall reach the packing room in faultless condition."[13]

Laundry owners often thought of laundry in production terms, measuring success in the aggregate. One lost tablecloth would not put a laundry in the red. Perfect service would not necessarily result in a profit. Furthermore, as laundry owners tried to shift their attention to the concerns of female consumers, gender differences added extra potential for misunderstanding. Laundrymen had direct experience with which to sympathize with the male customer who complained about the cutting pain of overstarched collars, but women seemed to have more mysterious standards for judging laundry work. In the laundryman's world, such mistakes were an inevitable part of the production process. They found it hard to sympathize with women who vested damage and loss with greater meaning.

Laundry journals sometimes used parody and criticism to create sympathy for customers' complaints. For instance, a British journal ran a series of mock letters from the fictional "Autocratic Laundry." One addressed the problem of laundry marks:

> Dear Madam,
>
> . . . We fail to see what you have to complain about. Entirely free of charge you have upon your underwear marks which would prove your identity should you at any time suffer from loss of memory or should you meet with an accident. If you send to us long enough, you will have your handkerchiefs completely bordered with distinctive marks–a border which will be ornamental and guaranteed fast dyed.[14]

The testimonies of female informants provided another means of creating sympathy among laundry owners. In an address before the Boston Laundryowners' Bureau, journalist Helen Hopkins recounted how poor laundry marketing and poor service had ruined her marriage and sent her to the hospital with a thyroid condition. The address was reprinted in *Laundry Age* and titled "Romance Dies beside the Wash Tubs." It was preceded by a note from the editor admonishing "Every laundryowner, every laundry advertiser, every laundry salesman in this country (and every other country where power laundries are operated) should read this story. Doing so won't be difficult for it is told with a frankness as revealing as a Broadway column."[15]

Laundry owners might have grown to sympathize with some of their women customers, but they also wanted protection from illegitimate claims. Articles offered advice on how to nip such complaints in the bud. One laundry owner found customers often filled lists out incorrectly, so he employed a woman to telephone customers whose lists did not tally with their bundles. Most, he said, admitted to their error. More revealing of the tension between demands of the process and customer service was his frank statement that he expected to pay for legitimate claims on about three-tenths of one percent of goods handled. In other words, he admitted that his laundry routinely ruined or lost three shirts or sheets out of each thousand. He also thought it good practice to use "extra" unclaimed items to round out bundles with missing items.[16]

Journals also offered advice on reaching and educating women consumers. Laundry owners looked for ways to take advantage of the gender-specific advertising strategies becoming common in this period. One enterprising Kansas City laundry ran a pseudo–advice column on the women's page of a local paper. In it, a fictional "Mary Steward" doled out advice on laundry and life. The Steward column ran directly underneath Dorothy Dix's popular syndicated column.[17]

Other laundry owners looked for ways to utilize women's patronage of radio and movies. In an article entitled "Cornered! They See, Hear, and BUY!" *Laundry Age* advised that "theatre patrons cannot escape your ads on local movie screens." Laundry owners could rent short films advertising laundry services from a central syndication agency to be shown before the feature film.[18] One Hollywood laundry went even farther in cashing in on the allure of movies. Twenty-nine "well-known" motion picture stars were made stockholders in the company. The laundry's grand opening featured Jackie Coogan, "the firm's youngest stockholder," demonstrating his ability to operate laundry machinery.[19]

The LNA ran an additional advertising and publicity campaign that opened up laundries for tours by women customers. In an American advertising campaign sponsored by the LNA, women's magazines such as *Good Housekeeping* featured the adventures of "Alice in Launderland." The text and pictures of a 1928 ad suggest the complexities of preserving the laundry owners' gendered authority while pleasing women customers. Alice, fashionably dressed in a cloche hat and fur-trimmed coat, is being shown through a laundry by a man described as her "genial guide." In the background are workers sorting laundry. "Here's our 'raw material' as it comes from the home," says the guide, using a factory analogy. It is the method, the technology, that is emphasized. Nothing is said about the women workers in the picture. Further text picks up another of the laundryman's themes: "Whether you supervise the laundress at home, or send clothes out to questionable quarters, you will find that modern laundries offer freedom from work and worry." And finally, a small box at the bottom of the page offers: "Go with Alice into Launderland–This delightful journey booklet may be had from any modern laundry displaying on its trucks this picture of Alice in Launderland."[20]

British laundry owners offered tours to women customers as well. One author of a British article thought issuing invitations to homes for such visits was a waste of time, however. He claimed that the public "never, or hardly ever" responded. "We doubt whether laundries who invite the inspection of their patrons ever get more than half a dozen visitors a year, and these visits usually take place when the customer has complaints to make." He thought such visitors were "not in a favourable state of mind to appreciate the beauties of the laundry."[21]

Throughout the interwar period, interviews with women consumers continued to reveal the same complaints and demands. Some laundry owners responded, some did not. More fundamentally, these social

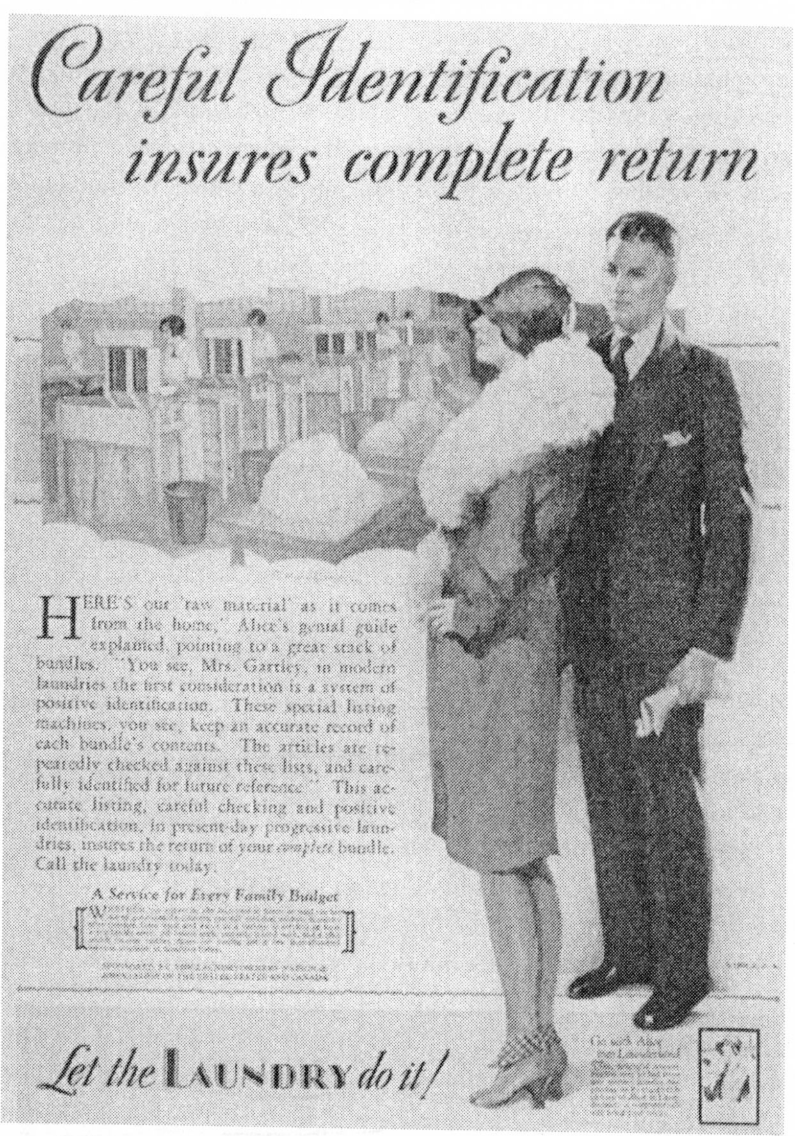

Fig. 18. *Good Housekeeping* magazine ran this advertisement for laundry services as part of a publicity campaign sponsored by the LNA. *Good Housekeeping* 87 (December 1928), 183. Courtesy of International Fabricare Institute.

measures could not solve some of the industry's basic problems. They could not eliminate the small margin of damage and loss that created a bad reputation for the industry as a whole. Truly incompetent laundry owners continued to do business. These advertisements and tours may, in fact, have raised expectations, making instances of bad laundry service even more damaging to the industry as a whole. On the other hand, laundry owners had always depended on a combination of the social and the technological to bring them business.

Looking for Henry Ford

In addition to customer relations, laundry owners looked for technological solutions to increase their customer base and to keep their old patrons during changing circumstances. Lowering labor costs was their most pressing problem. Despite the spread of ironing machines and other devices that mechanized finishing processes, laundry work remained extremely labor-intensive. Workers individually marked and sorted every item entering the laundry, counting and sorting again before it was packed at the end of the process. In between, each of those handkerchiefs and collars and pairs of underwear had to be individually shaken out after washing and fed through a mangle or pressed on some other kind of machine.

Old-timers often stated this kind of production exemplified a remarkable change from the earliest years of the century. On the silver anniversary of George V's coronation, in 1935, British laundryman J. J. Stark looked back to a time before the beginning of the king's reign, when 20 percent of the total employees in many laundries were hand ironers.[22]

American laundrymen did not measure progress against the reign of a long-lived king. Instead, they held up industry giant Henry Ford as a paragon against which they could gauge their technological and organization shortcomings. Silas Kidd, "the suds philosopher," shared with readers his wife's opinion that laundry prices were too high. "By gosh, if figures do not lie, my wife was justified to buy that double action wash machine with stainless steel and hoops of green," Silas intoned. He thought emulating Henry Ford was necessary: "If Henry were a laundryman he'd operate on another plan. I bet he'd get his prices down till every wash machine in town would gather cobwebs, dust and dirt." Kidd's prescription: "You've got to have expanding market and VOLUME is the only way to make the laundry business pay!"[23]

For well-established, full-service laundries, continually updating existing processes provided a hedge against changing public tastes and rising wages. Other innovations also helped lower costs. By the early 1920s, visitors to both British and American laundries could find conveyor belts widely in use. In these large laundries, drivers no longer needed to carry their load to the marking room (which precluded opportunities for flirting and sexual harassment–either a drawback or a benefit, depending on one's point of view). Tumble dryers also became widely available, replacing the onerous task of hanging garments in a drying closet. Some laundries hybridized old and new, running a conveyor belt with hooks on it into the closet.[24]

This kind of technological development proceeded incrementally. Flatwork ironers could be made to run faster and ironing machines redesigned to be less tiring for operators, but no machine fed itself or folded newly ironed sheets and napkins. Piecemeal improvements could not adequately compensate for the rising cost of labor or the lack of skilled operators. Because of the weekly laundry cycle, having a plant completely dedicated to fully finished work also left machinery, particularly washing machines and extractors, sitting idle at the end of the week. Laundry owners had no way to make use of this wasted capacity since, unlike manufacturers, they could not make an extra stock to be sold later.

The prices of fully finished service continued to rise faster than middle-class incomes. Over and over again, longtime customers complained that purchasing full services strained family budgets. Moreover, to expand their customer base, laundry owners needed to offer prices low enough to attract new customers, particularly apartment dwellers and lower-middle-class housewives. In an attempt to capture this market, an increasing number of laundries began offering a variety of inexpensive "alternative services" that eliminated marking, sorting, and ironing–labor-intensive processes carried out by relatively well-paid workers.

Bagwash, bobwash, and *wet-wash* were all variations on a basic concept. The customer would send all her (in the laundry owner's mind this customer was always female) washing in a bag. In many cases, the laundry was not sorted but rather dumped as a lot into a washing machine (often separated from other people's laundry by a net divider), put through an extractor, and perhaps a tumble dryer, and shoved back into a bag or box for the customer to iron and fold.

Within this basic framework, inventive laundry owners introduced a variety of gimmicks and variations. Customers who selected what was called in Britain the *bobwash* were provided with a net bag. They paid the laundry a "bob" (a shilling) to wash as much laundry as could be stuffed into the bag; the entire bag was placed in a washing machine. Clothes washed this way tended to return from the laundry still very wet and not very clean. Thrifty housewives also tended to stuff the bags full enough to strain the backs of deliverymen. Other laundry owners used newly available float ironers to partially press shirts, dresses, and underwear, which could then be ironed at home. Particularly in the United States, imaginative laundry owners seemed to be forever coming up with new variations. For instance, one Missouri laundry offered a "semi-finished shirt service." Workers starched and machine-ironed collars and cuffs. The rest of the shirt was left damp, to be ironed at home.[25]

In his 1935 retrospective, J. J. Stark credited this innovation to "those American pioneers who first realized the possibilities of conducting laundries on a mass production basis." He thought the Americans had been much quicker to realize the economic potential of such innovative, cheaper services, having introduced them as early as 1910. Alternative services did not catch on in Britain until the 1920s. By this date, such services were already widespread enough in the United States that the LNA had tried to standardize all services into six categories, ranging from the fully finished "prim-prest" to the least expensive "wet-wash."[26]

Alternative services depended on certain assumptions about the customer: that "she" considered the washing part of the laundry process burdensome enough to pay someone else to do it; that she would, however, willingly spend time ironing; and that she would tolerate the occasional accidents that came with washing unsorted laundry–such as leaking color or residual stains. In 1929, a British laundryman advised that alternative services "should be standardized, with no possibility of variation." The author rationalized his departure from traditional standards of service on both technological and social grounds. He argued that alternative services appealed to a lower class of customer. Laundry owners had previously given special service in order to create word-of-mouth advertising, particularly in the carriage trade. This laundryman believed such publicity would be more a nuisance than a boon when dealing with this new type of customer. Moreover, the success of alternative services would depend on "the 'mass production' methods of uniform treatment for all work."[27]

These services proved quite popular into the 1930s. Stark estimated that in Britain, alternative services grew from 7 percent of the total turnover in the industry in 1926 to more than 30 percent by 1933.[28] A survey of Philadelphia laundries during the same period found nearly 40 percent of customers opted for "semi-finished" services such as "damp wash" and "pak-a-bag." The survey also broke down statistics according to income level, confirming that families with incomes less than $2,000 a year most often took advantage of alternative services.[29] Laundry owners expressed more ambivalence about the economic viability of offering budget services exclusively. Many agreed they were most profitable when offered as one of a variety of options. Laundries used them to attract new customers who might later elect to splurge on more fully finished work, to keep old customers during hard times, and to keep machinery and workers busy.[30]

Laundries as Modern Businesses

Technological strategies were tightly intertwined with a broader set of approaches laundry owners sometimes called *modern business practices.* For laundry owners, being a modern business entailed employing a variety of techniques–careful accounting procedures, attention to markets and to the needs and desires of consumers, and the use of advertising.[31] Being "modern," or "progressive," also entailed a more nebulous set of cultural practices which allowed laundry owners to see themselves as the social equals of other businessmen, real and imagined. Employing progressive methods and participating in trade association activities not only helped make laundries more profitable but also allowed certain laundry owners to distinguish themselves from their more hapless colleagues. Laundry owners often claimed profit was the ultimate measure of the value of these techniques. In truth, laundry owners could not untangle economic strategies and rewards from cultural ones.

Members of the British trade publicly expressed ambivalence about this kind of progressivism more often than their American counterparts. Differing economic contexts certainly played a role. The series of recessions that hit Britain in the 1920s undermined laundry owners' optimism and expectations of progress and growth. In contrast, American businessmen (including laundry owners) enjoyed a probusiness culture during the 1920s, as well as the sense that the United States was rapidly becoming the leading economic power in the world. Even the shock of the 1929 stock market crash, more dramatically devastating to the United States than to Britain, did not entirely dampen this attitude.

The dominant theme in British business discourse of this period was intensive anxiety about international competition.[32] Despite the fact that laundries did not compete in international or even national markets, British laundry owners still adopted this language in comparing themselves to their American counterparts. In the early 1920s, Charles Roberts took a trip to the United States to observe American methods. He came back concluding that "American laundries were housed in much better buildings and were much better equipped than any I had seen in England, and they spent money on buildings, plans and machinery in a way which filled me with amazement."[33] Roberts stated that American laundry owners were able to do this because they received generous tax credits for depreciation—a comment that implied a critique of the British government's less sympathetic approach to business. British laundrymen also frequently claimed American laundry owners had enticed a larger proportion of the population to use laundries. British laundryman A. J. Frost summarized this sentiment in blunt, if not particularly logical terms, "We shall have to alter our ways to get ahead of them."[34]

British laundrymen also liked to suggest that Englishmen lacked a certain attitude characteristic of American businessmen, what the slang of the time called *pep* or *hustle*. Frost observed that "the temperament of the nation is different. We move slowly; the American is always out for new ideas, and is always ready to try a new machine."[35] After his American visit, J. J. Stark satirically noted the hidden benefits of the more laconic British attitude: "It certainly was a very inspiring thing to meet those men, old and young, who were always hurrying, until they went into the cemetery." Stark surmised that "some of them go rather earlier than we did in England."[36]

British trade journals also tended to claim progress in terms of contribution to the common good more often than their American counterparts. They saw technological change within laundries as improving both profits *and* working conditions. In the same spirit, Stark and others argued they provided a public service–now available even to the working class, thanks to a rising standard of living and the "bobwash."[37] In contrast, American journals most often measured progress only in terms of specific economic gains and technical changes.

In the British laundry community, Charles Roberts was recognized as an exemplar of the "progressive laundryman." He shared a number of characteristics with others of his kind on both sides of the Atlantic. For instance, he was always on the lookout for new machinery and new services that would benefit Millbay. At an annual conference at Brighton he heard an address on the "dry wash system." Upon returning home, he implemented the system in the family laundry, though it proved to be a financial failure. His later trip to America resulted in an order to the British-American Laundry Machinery Corporation for new, faster centrifugal extractors.[38]

In the British context, Roberts's antithesis was the laundry hobbyist, who, like an inept gentleman farmer, could poison the ground for others who came after him. An article in *The Power Laundry* offered a scathing portrayal of this pitiful creature, who, "having money on hand, generally sinks a thousand or two in a laundry" because "his aunt's cousin has a friend" who did very well in the business. The hobbyist begins with little experience of the laundry trade; he has "perhaps just walked through a laundry, but knows nothing about working one. Nevertheless he starts boldly and feels sure his mission on this earth is to teach the laundry people in his district a thing or two." For expert advice, he "confers with his wife and other female relatives, and between them they draw up a price list based on the charges they would like to pay for their laundry work." The inexperienced laundryman's "business is allowed to hum on till the end of his financial year, when the day of reckoning comes and his auditors give him the jerk of his life."[39]

The stereotype of the laundry hobbyist did not seem to be part of the American business bestiary. Instead American laundry owners characterized lack of professionalism in a variety of other ways. "A businessman is a businessman in any language," opined one author in an article entitled "Businessmen vs. Alleged Businessmen." Advocating the necessity of "hustle," he further ventured, "The only way to conduct a successful laundry is by being on the job, daily, from gong to gong . . . and, when I say on the job, I mean on the job and not in the back room of a kelly pool parlor."[40]

These kinds of articles became more common in the 1920s, as the trade literature began to make an increasingly sharp distinction between amateurs and professionals. Both American and British laundry owners imbibed the discussions of professionalization, ubiquitous during the interwar period, and they recognized the usefulness of the concept. Professionalization would contribute to standardization and would allow laundry owners to more effectively screen potential managers and other workers, raising the status of employees and employers. "Why not take a leaf from the book of the medical profession?" argued one American laundryman. "Years ago, in order to become a doctor, about all one had to do was to work for a doctor long enough to learn what he knew. You were then a doctor. If your patient had something that the other doctor never had to treat, then he died."[41]

Trade associations recognized that training programs and certification offered the most obvious technique for making the professional boundary more explicit. Training programs multiplied on both sides of the Atlantic after the turn of the century.[42] In practice, however, the medical model of professionalization was unworkable and undesirable. Laundry owners would have bitterly resisted licensing and formal training courses, which often did not offer enough rewards to make the investment worthwhile.[43] Like the erstwhile doctor, most laundrymen and laundrywomen learned the business by working for someone in the trade, and they would continue to do so. Professionalism remained mostly in the eye of the beholder.

Trade associations and trade journals continued to remain the center of the laundry community and the main impetus behind reform and modernization in the industry. Early associations had functioned primarily as a way to facilitate conventions and other interaction between members. By the 1920s, their role had expanded to encompass education, lobbying, research, and the setting of standards. In 1911, the LNA leadership realized that under the association's existing system of organization, it "could never become much more than a social affair." That year they substantially rewrote the bylaws so that each state organization that was affiliated with the national body would be obliged to contribute a portion of members' dues.[44] This money became the basis of the national association's expansion into research, education, and publicity programs. Sharing the common interest of representing the industry in negotiations with the British government over minimum wage rates, the two national British associations amalgamated in 1920, becoming the National Federation of Launderers.[45] Laundry owners, particularly in large cities, could also belong to a plethora of local associations, ranging in character from social clubs to highly active political organizations.[46]

The broader context for this zeal for organization and reorganization was the general growth of trade associations during the period. Laundry owners had been among the pioneer industries to form trade associations in the 1880s, but by the 1920s, the butcher, the baker, and the candlestick maker all could take advantage of the benefits of association.[47] Laundry owners could also join a variety of local and national organizations, including service clubs and the chamber of commerce (in the United States), where they might meet other businessmen.[48]

Making competition more predictable and stable was one of the longest standing functions of trade associations. In an industry that lived on its week-to-week receipts, and in which losses were never recoverable because laundry owners produced no "inventory" they could store and sell when the market became more amicable, price-cutting wars could spell disaster. Monopolistic agreements to set minimum prices for a given city or region provided an obvious solution to the danger of price-cutting.

In the 1920s, both American and British Associations would have stood on shaky legal ground had they created explicit price-fixing agreements with their members. Parliament held an inquiry into laundry profiteering in 1920, but it concluded that the industry was too idiosyncratic, too sensitive to customer demand, and too disorganized to effectively circumscribe effective competition.[49] In general, the British government was far less concerned with monopolistic practices than the federal government in the United States. Although the National Federation of Launderers only "recommended" price levels to its members, local associations regularly reached contractual agreements among members.[50] Various forms of government regulation, created with the input of trade associations, eventually proved at least a partial solution. Meanwhile, trade associations also tried to encourage laundry owners to learn the rudiments of accounting and business economics so that they would understand how to realistically price their services.[51]

Less coercively, by publicizing certain practices and techniques, trade associations hoped to encourage backwards laundry owners to conform to an industry-wide set of standards. "We know that a well conducted laundry is a better competitor than an inferior one. The latter brings the industry into disrepute and it suffers as a whole," stated an American laundry journal.[52] Another article put it differently: "Methods and quality require standardization so that the public may know what to expect. The customer finds, generally after numerous painful experiences, a laundry which suits her purpose. She moves to another town, and has the right to expect similar service."[53]

Trade associations had always promoted the idea of better practices. In the past, they had functioned primarily as a forum in which individual laundry owners could share their ideas. During the 1920s, the trend of forming such institutions in every field became part of a shift of emphasis towards a universal set of standards, most often those set forth by research bureaus. Pooling resources became a way for small-business owners to emulate the techniques of corporate giants like DuPont and Ford. A thousand laundry owners together could fund ongoing research and development or hire a Madison Avenue agency to design a national advertising campaign.

By the 1920s, the LNA had departments to research and provide information to its members on accounting and cost finding, engineering, advertising, and human relations. It funded the work of two chemists at the Mellon Institute of Industrial Research. Dr. H. G. Elledge and Alice L. Wakefield, a "graduate chemist," tested various soaps and other laundry chemicals, publishing manuals of standard practice for members of the association. The LNA's department of accounting and cost finding developed standard accounting procedures for laundries. The association also ran a high-profile advertising campaign in the *Saturday Evening Post* and other popular magazines.[54] Donations and membership fees paid for all these services.

The British Federation of Launderers also employed this model, but it was not as successful in developing the kinds of internal departments that characterized the LNA. The Federation did create a laboratory called the Scientific Research Association in the early 1920s. The laboratory was funded by a government grant (apparently a misguided attempt to improve British competitiveness) rather than by federation dues. Accusations were made that certain members of the federation kept the research data a secret, using it only for their own gain.[55] By 1935, the federation had managed to establish a permanent body, the British Laundry Research Association. In 1935, it was supported by five or six hundred laundry owners who together contributed nearly five thousand pounds a year for research.[56]

Progressive business practices, good publicity, and associations were not only tools for laundry owners to improve the economic viability of their businesses but also protections against the kind of status anxiety that came from being identified with a small-scale service industry catering primarily to housewives. As the previous discussion suggests, ideas about business practice in the laundry industry were not formed in a vacuum. Like other business owners, laundry owners read newspapers and magazines and talked to people in their communities.

The self-conscious did not need to look far to discover what others thought of them. They might have read with frustration *Babbitt*, Sinclair Lewis's 1922 novel of small-town business culture. In the novel,

George Babbitt, a small-time real estate agent, debates with his wife whether or not to invite the local laundry owner, Orville Jones, to a dinner party. Babbitt argues in favor of Jones:

"Orville is a mighty up-and-coming fellow!"

"Yes, I know, but–A laundry!"

"I'll admit a laundry hasn't got the class of poetry or real estate, but just the same, Orvy is mighty deep. Ever start him spieling about gardening?"[57]

A similar but subtler message could be read in the back-handed compliments of Portland's mayor as he welcomed the 1921 LNA convention to the City of Roses. According to *Laundry Age,* he used his opening remarks to observe that all the LNA members were "prosperous looking." In his opinion they were "the equal of business men generally in appearance. The banker, the merchant, the lumberman, the manufacturer have nothing on the laundry owner. . . . They hold their chins up high, are as proud of their business as other businessmen."[58]

For many American laundry owners, the LNA offered both a refuge and a solution to these kinds of problems. The editor of *Laundry Age* was quite explicit about this. The "chief aim" of the LNA, he wrote, "is to assist laundryowners in raising the general standing of the industry to the high place of other important and essential industries."[59] "Let's stop apologizing," argued another article, stating that laundry owners are "apologetic where they have every right to be proud, obsequious when they should display righteous indignation."[60]

One particularly cynical British observer argued that highly organized conventions, with their plethora of technical papers and self-congratulatory speeches, in fact wasted time and money, serving mostly to make participants feel superior to industry predecessors. He claimed, "The modern conference is not always much better than the old public house gathering, except that it has high brow atmosphere." Members discuss "atmospheric humidity instead of washhouse steam."[61]

British laundry owners often defined their status problems in more specific terms than their American counterparts. They saw the industry's standing as being undermined by poorly run businesses and negative publicity about labor practices. To gain status, they were not above manipulating the British monarchical status machinery to their own ends–an option not open to Americans. In the 1930s, industry leaders decided the National Federation of Launderers should be renamed *The Institution of British Launderers,* thus making it eligible for a Royal Charter. A British laundryman explained to an American audience that he ultimately hoped "to see a launderer or two mentioned in the Honours Lists of the King's Birthday. A K.B. or at least an O.B.E."[62]

The presence of women managers gradually transformed the social dynamics and internal culture of British associations, making them determinedly heterosocial. After the war, these women began to publicly question not only their right to attend meetings but also the dominance of men in the leadership of the federation and local associations. In 1919, *The Power Laundry* reprinted an address given by a woman laundry manager at a conference at Harrogate. She argued that associations would benefit from the expertise of women such as herself. "May I suggest a fuller and closer association, by experimentally calling the experienced women of our trade into our public administrative life?" she provocatively asked her audience.[63] British women indeed ended up finding a place for themselves in alternative institutions created within the federation. In 1924, the Young Launderers Association was created, and women quickly came to dominate within its ranks.[64] Some women also created their own institutions, such as the Women Launderers' Social Club. By the late 1930s, the federation had its first woman president.[65]

Some British laundrymen used social conventions such as the "toast to the ladies" to express hostility towards the women in the room. One such set of remarks was quoted in a trade journal under the title "A Merry Party at Grange-over-Sands." The author explained that in his remarks, T. H. Corbett conflated the wives in the room with the women managers. He told the assemblage that "the ladies took a very great interest in their business, and he could also say a lot about their influence on our lives. Who had not felt the balm of a cool hand on his fevered brow, the heavy corrective hand after an exhausting and strenuous time at his club or lodge, and seen the eager open hand on pay day?" Addressing the men in the audience, he said that "he felt sure they would agree with him that the ladies were essential, and particularly to the laundry business, but he wished they would go back in their fashions twenty years." A Mrs. Farrell, described as the wife of the section chairman, was reported as responding by suggesting that she "preferred the word 'women' to that of 'ladies.'" She noted that "this last few years had been an epoch-making time for women, but they had not quite got emancipation yet."[66]

The few female managers and owners in American associations remained largely invisible. Throughout the 1920s, the *National Laundry Journal* ran a column entitled "About Men in the Industry," which faithfully printed two-paragraph profiles about the hobbies and business accomplishments of laundrymen. Women did attend regional and national conventions but were often excused from the room after the opening address. Trade journals commented on their presence mostly in terms of their decorativeness and the quality of moral uplift they brought to the proceedings.[67] The "new woman" might have been making inroads on business, but American laundry owners did their best to ignore her unless she had something useful to tell them about what women consumers wanted.

✖

During the 1920s, laundry owners invented strategies to attract new customers and keep old ones. Hampered by the technological limitations of their businesses, they looked both to new services that would more efficiently utilize machinery and marketing techniques that would improve public perception of the industry. Through trade associations, they also sought out methods to improve the internal organization of the industry and to raise their own social standing in the community.

The results of these efforts were mixed. Meanwhile, pressure from without and within was continuing to build, fomented by workers and reformers. Global economic depression and the rise of new consumer cultures would further add to their challenges.

Laundries in Black and White

THE INTERWAR PERIOD WAS MARKED BY A GRADUAL DEMOGRAPHIC shift in the workforce of American laundries, as a growing number of African American women (and African American men, in fewer numbers) began to replace the European immigrants and native-born white women who had long monopolized laundry jobs. This change altered the industry in a multitude of ways. On the shopfloor, it created a two-tiered workforce primarily divided along racial lines. Wages, division of labor, and workplace culture were all affected. Race also became a central factor in efforts to create unions and to reform the industry through legislative means. This American pattern had no equivalent in British laundries, which still remained racially and culturally homogeneous. Racial differences in the British context functioned only on the level of symbol and ideology, not in creating day-to-day relationships.

It has long been recognized that World War I became a watershed in the entry of African Americans into factory work in Northern cities. The massive migration of Southern blacks, the increasing unwillingness of whites to take the lowest paid and most difficult factory jobs, and the severe limits placed on immigration after 1924 caused a growing number of employers to look to African Americans to fill jobs once claimed by other groups.[1] Laundries were no exception. In 1919, the Chicago Commission on Race Relations found that "the opportunity to work in a laundry was practically denied to negro women until labor shortages forced laundryowners to tap this reserve labor supply."[2]

Black women in particular flocked to laundries as an alternative to domestic service and Southern agricultural work. The Chicago Commission stated that "Negro women were eager to desert work as domestic servants and 'family' washer-women with the social stigma and restricted human contact involved." Though hot and difficult, commercial laundry work was still more appealing than its domestic analogue. Like other kinds of factory work, "more independence is possible, hours were better, and association with fellow workers enlivened the work day."[3]

After the turn of the century, census takers began tracking these changes, with varying degrees of accuracy. In 1910, they found that 16 percent of the women who identified themselves as "laundry operatives" (as opposed to washerwomen) were African American. By 1920, the figure had risen to 26 percent. And although black workers were the first to lose their jobs after the 1929 stock market crash, the percentage still continued to increase through the 1930s.[4]

These statistics disguise regional variations. Laundry work remained an occupation primarily for white women in the West, in Boston, and in the smaller cities of the Midwest. Because many white women in the South would not work alongside African Americans or take jobs identified as "colored," black women often constituted 80 to 90 percent of the workforce in Southern urban laundries. This pattern was further reinforced by the extremely low wages that Southern laundry owners justified on racial grounds.

The biggest change in the racial demographics of laundries was, of course, in those cities most affected by the northward labor migrations. Detroit, Cleveland, New York, and Philadelphia all experienced an influx of Southern migrants seeking jobs. Contemporaries often held up Chicago as a particularly dramatic example. In 1910, African Americans comprised only a tiny percentage of Chicago's laundry operatives. By 1930, census takers identified 55 percent as "negro."

Laundry work was extremely important to both recent Southern migrants and longtime resident urban African American women in making the transition from domestic service to factory work. Agricultural workers and domestic servants often found their first factory jobs in laundries. Between 1910 and 1920, commercial laundries experienced an increase in percentage of African American workers larger than that in any occupation except barbers and hairdressers.[5] Of 536 black laundry operatives surveyed by the Women's Bureau in the late 1920s, a little less than half (233, or 43.5%) listed domestic service as their only previous employment. The next largest category, 26 percent, had worked only in laundries.[6] In contrast, few white laundry workers had previously been employed in domestic service. The largest group of white respondents (33.4%) had spent their whole work lives in laundries. The next largest group had previously found employment in factory work. While only 6 percent of the black women had ever worked in other types of factory work, 21 percent of the white women had.[7]

Laundry work had long been an important source of employment for female newcomers to wage labor. However, immigrant women had spread themselves out over a wider range of industries. A far larger percentage of black women spent time working in laundries than did their white, immigrant predecessors. Why did these African American women choose to seek work mainly in laundries and why were employers willing to hire them? While it is clear that fewer and fewer white women wanted laundry jobs, leaving a hole in the labor market, black women did not simply fall into that hole by taking jobs no one else wanted. They actively sought laundry work rather than other kinds of employment (some of which was even more unacceptable to white women) and were hired in large numbers by employers.

One supplementary explanation of why so many black women entered the industrial labor market through laundries lies in the identification of laundry work with domesticity. Although mechanization made commercial laundry work very different from taking in washing, the domestic connection may have helped convince employers to hire African American women. Well into the twentieth century, the black washerwoman was not only a pervasive cultural stereotype but also a social reality. In both the North and the South, census takers found an extraordinary number of African American women who identified themselves as washerwomen. In 1920, almost 20 percent of all African American women surveyed identified themselves as "laundresses (not in laundry)."[8] Even in Northern states, African American washerwomen outnumbered their sisters working in laundries by almost two

Fig. 19. Many African American women from the South began their work lives by doing wash "on the yard." Photograph by Jack Delano, 1941. Farm Security Administration Collection, Library of Congress Prints and Photographs.

to one. In the South, this gap was much larger. In relatively urbanized Georgia, 36,775 black women earned their living taking in washing, while only 2,094 worked in laundries.[9]

Previous experience with doing family wash may also have influenced black women to look for work in a laundry rather than in productive industries. Most Southern migrants either had experience themselves or had helped female relatives working "on the yard." Sylvia Woods, an African American laundry worker who would later become a labor organizer and activist, got her first job in a laundry around 1924. A newly arrived Southern migrant to Chicago who had never had a job, she returned each morning to a neighborhood laundry until the manager relented and hired her. Apparently, she never thought to look elsewhere. Though she did not specify why she went to a laundry, she did explain that her mother had washed for a white family in New Orleans.[10]

Once hired, African American laundry workers entered into a complex web of relationships shaped by race, class, and gender. These relationships dictated what jobs they could work, how much they would get paid, and how they would relate to coworkers and employers.

As a group, African American women earned the lowest wages among laundry workers. However, there were huge regional differences in the disparities. The larger the percentage of African American

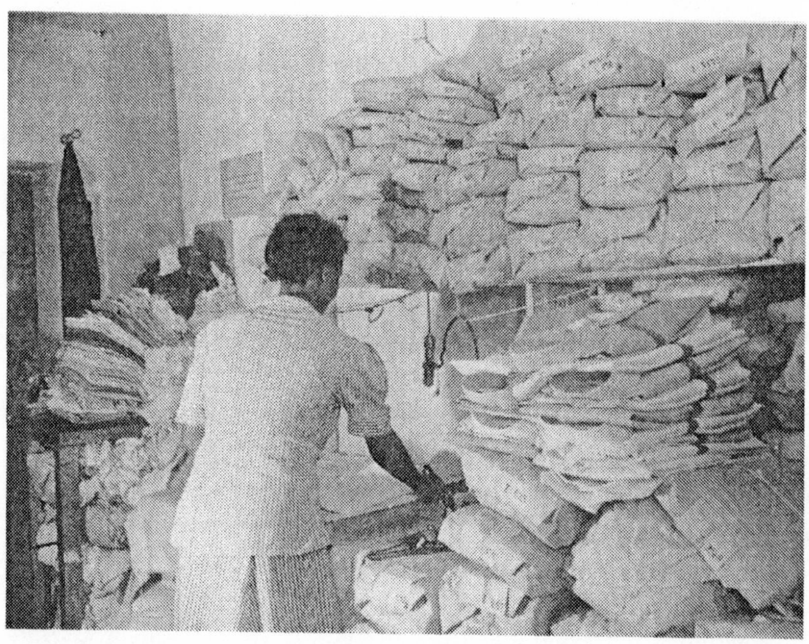

Fig. 20. During the interwar period, even some Asian American laundry owners hired African American women as workers. Farm Security Administration Photographer Gordon Parks entitled this 1942 photo "Johnnie Lew's Chinese Laundry on Monday Morning." Farm Security Administration Collection, Library of Congress Prints and Photographs.

women and the deeper into the South, the greater the wage gap. The 1927 Women's Bureau survey found that the median income of black women in the Eastern United States was 88 percent of what it was for white women. In the Midwest, this figure was 83 percent, and in the South, 53 percent.[11] Investigators who resurveyed this workforce in 1934 found similar regional differences. They also discovered that the few black laundry workers in Los Angeles actually earned slightly more than their white counterparts.[12]

A variety of factors contributed to these differences in earnings. Both Northern and Southern laundries preserved categories of skill and the wage differentials that went along with them. African American women were most likely to work as shakers or machine operatives. Previous domestic experience also gave them some access to the remaining hand ironing jobs. Women's Bureau investigators found black women in every department of Northern laundries–including checking and marking, the jobs with the highest status. In Southern laundries, however, African American women were more restricted. They almost never had access to high-status jobs. As with their immigrant predecessors, African American women seldom found jobs in the "carriage trade" laundries. In the South, these laundry workers were most likely to be found in the lowest-paid segments of the industry, notably towel and linen supply houses.

In addition, Southern racial codes allowed laundry owners in cities like Atlanta, Birmingham, and Houston to be quite inflexible about the race-typing of jobs. Although Northern laundry owners also gave out jobs partly on a racial basis, they were more willing to reward talent and the ability to negotiate higher wages. In some cases, higher wages in the North can also be accounted for by the minimum wage laws in that region that came and went during this period.[13]

Patterns of employment further affected earnings. Black women worked more hours, but they were also laid off more frequently. While black women could be found in every position at a laundry except that of supervisor, they were more likely to be relegated to shaking and operating machines, while white women worked as markers, sorters, and fancy ironers. This was particularly true in the South, where white women monopolized marking and sorting jobs.[14]

While these statistics illustrate general employment and earning patterns, they can only indirectly reveal the attitudes and experiences of individual workers and employers. In general, historians have done very little work on race relations between women workers or between African American women and their white employers.[15] We do not know whether gender mitigated the virulent antagonism that often characterized such relations between male workers, or whether racial distinctions had different meanings and functions for working-class women than for men or for employers.

Not enough evidence survives to use the example of laundry workers as a definite case study for larger race relation trends. However, anecdotal evidence does suggest the complexity of shopfloor race relations, particularly in the North. While the attitudes and policies of Southern employers towards African American workers were often predictable, in the North the patterns created by an infusion of these new workers were novel to both workers and employers. In 1910, African American workers were mysterious enough to Northern laundry owners that the *National Laundry Journal* published an article entitled "How to Handle Negro Help in the Steam Laundry?" offering advice to prospective employers.[16]

One important factor influencing race relations in laundries was the attitudes of individual managers and laundry owners. Some employers saw race as a tool for creating divisions among workers. An African American woman told one investigator that "a colored girl would come in and the boss said 'we can't take you on. The white girls don't want to work with you.' The colored girl would start to go and he'd say 'I'll take you but you got to work for less money than the white girls.'" This employer would then fire white workers and replace them with black ones, further deepening race antagonism between the workers. African American workers also frequently complained that they remained stuck in the lowest-paid jobs while promotions invariably went to white women.[17]

The experiences with white workers described by Sylvia Woods further illustrate the complexity of shopfloor race relations. At the laundry where Woods first found employment doing flatwork ironing, jobs were generally divided along racial lines. White women fed the mangle, black women shook out the work. Woods got the white women to teach her how to feed and fold–giving her the skills necessary for access to higher paying "white jobs." She also asked how much her white coworkers were being paid and used this knowledge to ask for a raise. A white worker warned her not to: "Kiddo, don't do it because you'll be fired if you do." Woods got the raise but was warned not to tell the other workers. When she defied him, the boss relegated her to the night shift, which had an all-black workforce. When he then discovered that these workers were stealing sheets as revenge for their lower wages, the boss tried to impose a white woman supervisor. The black workers refused to work for her.[18] Woods's story is interesting in part because of its suggestion of interracial cooperation as well as the apparent fluidity between "black" and "white" jobs.

African American women from rural or domestic service backgrounds had to learn the ways of industrial work, just like the immigrant workers that had come before them. Employers made it clear that they preferred women born in the North to new migrants because these local women already knew what was expected of them. Not only were machines and schedules unfamiliar to Southern migrants, but such workers also brought with them a culture of resistance they had developed to deal with white employers in the South.[19] Accustomed to extraordinarily low wages and horrendous working conditions, these workers learned to pace themselves. Southern employers tolerated such behavior, which they rationalized as "natural" to blacks and even used to justify the low wages they paid black workers. The cycle of racial prejudice and low wages in the Southern labor system created a vicious cycle that reinforced both employer attitudes and worker resistance.[20]

Many Northern employers had no compunction about using whites and blacks interchangeably. They expected a certain level of performance from all workers and simply fired those they judged unproductive. Experienced Northern workers understood this. They also understood that, unlike Southern employers, many Northern laundry owners would reward a skilled, hard-working woman with relatively high wages–even if she was black. Workers like Sylvia Woods therefore employed other means of resistance and negotiation that helped achieve their goals better than foot-dragging and soldiering.

These shifting employment patterns did not escape the attention of the reformers, who had long interested themselves in laundry workers specifically and women workers more generally. Members of organizations created in the progressive period, notably the Women's Trade Union League and the Consumers' League, as well as new state and federal agencies such as the Women's Bureau of the United States Department of Labor, all made African American women and race relations an explicit part of their reform agenda during the interwar period.

These women reformers sought advice from and made alliances with various civil rights groups. Together they tried to make a place for black women in the labor movement. Less successfully, they also argued against the racial discrimination in New Deal labor and relief programs. Although their efforts extended to a variety of workers, these reformers made a special effort on behalf of laundry workers. Like their predecessors, they treated the problems and successes of laundry workers as symptomatic of broader patterns in industrial America. During the interwar period, those issues included not just class and gender but also race relations.

As migration from the South increased after the turn of the century, women involved in social work, settlement houses, and labor organizing could not help but be aware of the special needs and problems of African American women. Jane Adams and a number of other prominent progressive women were early supporters of the NAACP, founded in 1909, and the National Urban League, the principal organization for studying and helping urban blacks in the interwar period. During World War I, women in the labor community gained further experience through war mobilization, which brought many black women into factory work for the first time.[21] Immediately after the war, the Consumers' League, the Women's Trade Union League, and the newly founded Women's Bureau all began specific efforts aimed at understanding and ameliorating these women's circumstances.

Most of the reformers whose work was directed at laundry employees had a direct or indirect connection with the WTUL and its New York affiliate (NYWTUL). As previously noted, the NYWTUL and its national parent organization had long tried to organize laundry workers and had been important in the push for protective legislation. By the 1920s, the leadership of the WTUL had begun to pass from elite women into the hands of working-class women. The laundry worker's story is particularly intertwined with the career of Rose Schneiderman, the former garment worker who became president of the NYWTUL in the 1920s and was the only female member of the National Industrial Recovery Administration Labor Advisory Board in 1933.

During the 1920s, the NYWTUL cooperated with the Urban League and a variety of other African American labor organizers to unionize laundry workers. Through these interactions, Schneiderman and a succession of league organizers and organizations learned about the black community. By the 1930s, these efforts had become so normalized that race was no longer a matter of explicit discussion in the league. At the same time, the leadership of the African American labor community, with whom women reformers formed such alliances, seems to have been almost exclusively male. Part of the appeal to these groups of organizing laundry workers were the benefits to African American men, who made up a large proportion of the washroom workforce in towel and linen laundries.

In some ways, women reformers' long experience with organizing women workers may also have given them a greater understanding of black workers. Many male labor leaders were extremely impatient with workers who had little knowledge of what they termed "labor union principles." They tended to be dismissive of workers who could not fit into trade union culture, who did not understand the rituals and the language of this essentially white, male craft world. These women understood the troubles of workers who had not had the experience of apprenticeship. Many came out of rural backgrounds themselves or had previously worked only in domestic service. In classes and in meetings, the league taught these workers the forms and behaviors of labor culture. From their experience with immigrant women, Schneiderman and others also understood that organizational efforts among workers outside the white, male mainstream had to be reshaped to be appealing to these workers. The price for not doing so would be empty meeting halls and failed unions.

The Women's Bureau of the Department of Labor was also extremely important in providing reformers with information about laundry workers and in influencing federal policy during the interwar years. The league also became a nexus for a succession of women who would be important in making federal policy. From the organization's very beginning, at the end of World War I, the leadership of the Women's Bureau took a special interest both in African American women and in laundry workers. In 1919, Mary Anderson, director of the Women's Bureau, hired Helen Brooks Irvin, an African American economist, to advise the Women's Bureau on the status and needs of African American women workers.[22] They helped develop in the Bureau both a special interest in laundry workers and in African American women. This consciousness towards race is also apparent in the way Women's Bureau reports created categories of analysis. Whereas most progressive period reports divided workers into ethnic categories and "foreign born" vs. "native born," by the 1930s most workers were categorized only by their sex and race.

To understand the actions of both women reformers and civil rights groups, it is important to understand the limits that each put on their activism. Many white reformers unconsciously brought to their work the racial attitudes of their culture. For instance, in reports and other publications they used racial stereotypes to add interest to their portrayals of workers. In an article on laundries for the WTUL journal *Life and Labor,* a white college student describes her experiences masquerading as a laundry worker. She is careful to point out racial inequalities in wages, but she also repeatedly describes one of her coworkers as an "old Mammy."[23]

Women reformers understood that black women were part of their constituency, but they chose to prioritize gender over race. When political choices had to be made, they abandoned an insistence on racial equality first.[24]

⁂

In some ways, the influx of African American workers into the laundry industry mirrored the way earlier groups of immigrant women had used laundry work as an entrée into other types of factory labor. However, race had different meanings than ethnicity in the United States. Those meanings were further fragmented along regional lines and mutated by the relations of class and gender. Race would come to have an unexpected importance in shaping efforts to regulate and unionize during the interwar era.

Bringing in the State

IN 1948, CECIL ROBERTS WROTE A HISTORY OF HIS FAMILY'S PLYMOUTH laundry for private publication. In it, he recounted returning from the battlefields of World War I to find the once quiet, predictable haven of the Millbay beset with troubles. Economic depression, labor unrest, and changing consumer practices threatened the laundry's business. In addition, the British government notified Roberts that the laundry industry would soon be subject to a new form of regulation. Parliament had just expanded the scope of the Trade Board Acts, obliging Millbay to comply with regulations as of November 1919. Roberts claimed the act effectively doubled the wages of his family's female employees.[1]

Millbay survived the Trade Board Acts, the depression, and, most tenuously, nearly fifty German air raids that obliterated the center of Plymouth and burned part of Millbay's laundry plant. Roberts became a prominent figure in the industry, active in trade associations and an enthusiastic visitor to foreign laundries. It is therefore not surprising that he took a conciliatory line in assessing the long-term effects of state-mandated minimum wages. Echoing the trade association party line, he wrote that although the Trade Board Acts "caused a great deal of difficulty at the time of application it was becoming generally recognized that the industry had benefitted in the long run."[2]

When confronted by the possibility of extensive state regulation of prices and wages as part of the 1933 National Industrial Recovery Act (NIRA), many American laundry owners initially welcomed state efforts to stabilize the industry by regulating competition. This spirit of cooperation was short-lived. A number of factors outside of Roberts's experiences in Britain complicated American efforts. The relationship between wages and race proved particularly contentious. Many Southern laundry owners claimed they were not categorically opposed to some form of minimum wage, but they thought the $14-a-week minimum proposed by a Washington-based board was too much to pay their predominately African American workforce.

Both the NIRA and the Trade Board Acts departed from earlier forms of legislation by focusing on the previously sacrosanct area of wages. For an industry in which payroll constituted nearly 50 percent of costs, mandating minimum wages had profound implications.[3] By guaranteeing a minimum wage to even the youngest, least-skilled worker, these laws undermined the strategies of mechanization and the advantages of relying on cheap female labor. In return, however, the British government promised employers that the law would eliminate unscrupulous competitors. In America, the National Recovery Administration (NRA) promised to compensate businesses by setting minimum prices for services. This provided cold comfort to laundry owners, who believed their main competition was unregulatable housewives and washerwomen.

These seeming victories for workers hid a more complex side of regulation. The NIRA and the Trade Board Acts offered both business and labor an opportunity to reshape the industry through legislation.[4] They also gave a great deal of the responsibility for creating policy to interested parties rather than vesting it in experts, as had previously been done. These programs strove to obtain consensus from the affected parties by opening up policymaking to workers, employers, and the public through the creation of advisory boards.[5]

This strategy intentionally offered something for everyone: higher wages for workers, greater stability and some protection from unscrupulous competitors for employers. Compromise did not result in the outright victory of large business interests, however, as may have been the case in other industries. On the other hand, in both Britain and the United States, existing distinctions based on race, class, and gender were built into the way the law was written and enforced, undermining its more radical potentials for workers.[6]

The imposition of these programs had real consequences for the industry. Though trade association propaganda sometimes liked to blame the decline of the industry on state regulation, there is no convincing evidence that this was the case. Instead, the repercussions were more subtle, affecting laundry owners' choices about technology and labor force; determining which businesses would survive; and, most significantly, influencing the ability of laundry workers to organize. British laundry owners generally agreed that the Trade Board Acts were the most significant pieces of legislation affecting the industry in the interwar period. The legacy of the NRA was more ambiguous, but it definitely opened the door for effective labor organization among laundry workers.

For historians familiar with social policy history on either side of the Atlantic, the NIRA and the Trade Boards might initially seem like an odd comparison. Created in 1909, the Trade Board system became the basis of an industry-by-industry minimum wage system that still exists today. In contrast, New Dealers created the National Industrial Recovery Act as an emergency effort to stabilize an American economy devastated by the Great Depression. In their scheme, American industries would be organized into semiprivate regulatory organizations, empowered to set binding industry-wide standards not only for wages but also for prices. Although it paved the way for the revolution in labor laws of the "second New Deal," the NRA was, in and of itself, short-lived and ultimately unsuccessful, surviving only from 1933 to 1935.[7]

Despite these differences, efforts to regulate the laundry industry through the NRA and the Trade Board system involved elements worth comparing. Both were national in scope. The Trade Boards continued a longstanding British practice, while in America the NRA represented the first extensive effort to promulgate labor standards and other forms of regulation through the federal government. Although the NRA was intended to regulate not only labor standards but also price levels and conditions of competition, minimum wage rates became the central issue in shaping the laundry code. Both encouraged, directly or indirectly, employers and workers to organize themselves.

This resemblance between the Trade Boards and the NRA was not accidental. Both represented responses to a transatlantic conversation about how to ameliorate the problems of industrial capitalism in a way that was "democratic" and did not interfere with national economic well-being. Like many of the reform efforts described earlier, the flow of ideas was primarily from east to west.[8] In its structure and many of its goals, the NRA could be seen as a grandiose effort to emulate the European models with which many of its creators and functionaries were intimately familiar.[9]

The Trade Boards

The British Trade Board system originated with an act of Parliament in 1909. This act specified that minimum wage rates should be set on an industry-by-industry basis in the "sweated," or very low-wage, trades.[10] The actual rates were to be determined by representatives of workers and employers who, together with representatives of the public, formed the board. Once confirmed by Parliament, the rates would carry the force of law and would be enforced by representatives of the Ministry of Labour.[11]

The Trade Board Acts did not initially cover the laundry industry. However, many of the reformers involved with the initial passage of the acts had a longstanding interest in laundry work. Lady inspectors and representatives of various women's reform organizations lobbied vigorously for the industry's early inclusion, but Ministry of Labour staff members responsible for amassing evidence for such an inclusion soon discovered the difficulty of devising an argument that laundry workers were universally low paid. Their investigations revealed what others in the trade already knew–compared with other women workers, hand ironers earned relatively good wages and could not really be classified as "sweated labor."[12]

In 1913, representatives of the Ministry of Labour presented Parliament with an alternative measure. They proposed that wage levels be set only for calender (flatwork) and ironing machine operatives–the growing groups of low-paid workers created by mechanization. Because of some peculiar twists of legal reasoning and technological confusion, the original proposed extension of the acts excluded hand laundries and laundries using electrical (rather than steam) power to drive machinery. British laundry owners vehemently opposed the proposed statute. They argued, as did some of their American counterparts in the 1930s, that the government's actions would unfairly benefit their competitors–in this case, unmechanized laundries. The House of Commons sympathized with this argument and refused the extension in 1913.[13] There the problem would rest until the end of the war.

After the armistice, agents of reform again eagerly worked to expand the reaches of the acts. In 1918, the scope of the law was changed to encompass industries in which workers were poorly organized as well as those in which they were poorly paid. Reformers hoped that inclusion on these grounds would not only remedy the inability of these workers to negotiate collectively but also keep strikes from breaking out. Within this definition, the laundry industry appeared to be an obvious candidate for inclusion. This time, the writers of the bill to include laundries also made sure that the Laundry Act would apply to all but the smallest of laundries, in order to satisfy steam laundry owners' concerns about competition–an arrangement that ultimately benefitted larger, more mechanized laundries in which labor was a smaller proportion of overall costs.[14] Although coverage under the acts was supposedly not gender-based (unlike American minimum-wage laws of the same period), it nevertheless initially applied only to female workers. By 1920, however, revised statutes encompassed most male workers.[15]

Laundries were included under the Trade Board Acts at a time when employers and workers were under particular stresses. While Britain never experienced the all-out sense of crisis characteristic of the Great Depression in the United States, the end of the First World War was still marked by a sharp economic downturn (or "slump," as it was called). A lingering war tax further burdened the laundry industry, keeping prices up. One of the trade journals parodied the situation by reporting that washing had been seen hanging outside some of the grand houses on Park Lane, one of the most exclusive neighborhoods in London, as even the wealthiest consumers responded to high prices and hard times.[16]

The end of the war brought workers their own set of concerns. Many women war workers were pushed out of well-paying jobs after munitions production came to a halt and male workers returned. These displaced women flooded back into "women's jobs" such as laundry work, desperate for employment but, at the same time, newly politicized by their wartime experiences. Making this transition more painful, the government instituted a policy denying unemployment payments to women who turned down any job, no matter how unsuitable or low-paid—even those in the dreaded field of domestic service.[17] Unions, including the Federation of Women Workers, which had actively organized laundry workers since before the war, found a host of eager recruits. In the immediate aftermath of the armistice, the laundry industry was plagued by a wave of strikes carried out by federation members.[18]

On February 5, 1919, the laundry board had its first conference, beginning a series of meetings that would continue over the years. All boards consisted of ten to fifteen representatives meeting in closed sessions. Each side presented its argument and then proceeded to bargain for concessions. The intimate atmosphere of these meetings was one of the characteristics that distinguished the Trade Boards from the theatrical atmosphere of the NRA public hearings.

Proponents of both the NRA and the Trade Boards like to portray the meetings themselves as democratic forums where ordinary workers and employers (and consumers, in the American case) met to reach a common understanding. British reformer Dorothy Sells described to an American audience a homeworker from Dorset "who spent most of her days sitting on the cottage steps mending nets and looking out upon the changeless sea." Sells went so far as to say that when this net mender goes to London to attend the Trade Board meeting, "she learns something about the net making business, something about human beings, and about cooperative action that she carries back to her neighbors who sit on other Dorset steps."[19]

Sells exaggerated the role played by ordinary workers and employers. The acts called for "representation" of employers, employees, and the public. In practice, this only meant trade union leaders and the owners of large laundries who ran the trade associations. On the laundry board, directors of the National Federation of Co-operative Laundries and the National Federation of Laundry Associations represented the "employer's side." Representatives from six different unions spoke for workers. These included two powerful general unions that welcomed women workers: the Amalgamated Union of Co-operative and Commercial Employees and Allied Workers, as well as the National Federation of Women Workers.[20] Because of the British pattern of transindustry unions, a number of the labor representatives were not laundry workers at all but lifelong organizers, many of them middle-class or elite women reformers. The "public" was not intended to represent consumers, as in America, but rather the enlightened middle class, working under the principles of Fabian socialism. Sells described these representatives as typically "one lawyer, one professor of economics or sociology, and one lady distinguished for benevolent works."[21] In the case of the laundry board, many of the representatives held their positions for more than twenty years.

After extended deliberation, the Board produced a wage scale to be handed over to the Ministry of Labour for distribution to the industry. The "minimum wage," as set out by the board, bore no resemblance to the simplicity of the minimum wage system that Americans are now familiar with. It was minimal in the sense that it represented the lowest wages employers could legally pay their workers without special exemption.[22] The British Trade Boards set wages for every job in the laundry, from shaking out sheets to stoking the boiler. Most minimums were based on hourly wages, though inspectors developed piece-rate scales as well. The law exempted some categories of workers, such as learners and the aged. The board also made two regional exemptions: Northern Scotland, reeling under persistent poverty, and Cornwall, where wages were depressed from the collapse of the tin mining industry.[23]

Both the structure of the wage list and its interpretation by inspectors reflect the complex cultural and economic meanings of waged labor in British society.  The existence of the boards was predicated upon the belief that wage levels could be set to reflect what a worker was worth to an employer while also providing that worker with a living wage. Both British and American minimum-wage legislation also assumed the state could not set wages at a level that would prove ruinous to business owners. Official rates supposedly reflected wages paid by the "good" or "best" employers rather than an abstract formula created by civil servants.

Such reasoning was a convenient fiction in an industry that varied enormously in scale, degree of mechanization, and division of labor from business to business. In figuring such norms, the NRA and the Trade Boards relied on previous practice. In both cases, policy wavered between setting wages according to industry averages or according to the practice of "good" employers–which often meant the owners of large laundries who, owing to the volume of their business, dominated trade associations and could afford a higher wage scale. The British were more likely to act as if the two practices in wage setting were one and the same, while for many Americans the difference came to encapsulate fundamental questions about the purpose of the law.

While eventual responsibility would fall to organizations composing the board, the government needed a set of baseline figures to determine which workers should be included and to keep negotiating parties honest in their assertions about the industry. The Ministry of Labour lacked the statistical and census data available to Americans, which greatly hampered its efforts. In the initial 1914 attempt to get laundries included under the acts, the Ministry had resorted to sending an inspector to six London labor exchanges to collect figures on wage rates. The best she could come up with were figures from 1906. She argued that little had changed in the industry from 1906 to 1914, but no one was fooled by this argument.[24]

After the war, members of the Laundry Trade Board still lacked adequate statistics from which to figure rates. Instead, the 1919 rates reflected a scale arrived at by a board of arbitration, after a series of strikes sponsored by the Federation of Women Workers.[25] The threat of further strikes may have helped to push through such a compromise. FWW representative Mary MacArthur introduced a subtle threat at the first meeting of the Board: "As far as the women I represent are concerned they are determined to have their wages fixed by the Trade Board or some similar method." She then reminded the employers' representatives that a "stoppage of work" existed as an alternative.[26]

Despite these problems, members of the Laundry Trade Board were, in fact, able to agree and to convince most employers of the inherent fairness of their scheme. They paid careful attention to the construction of preexisting categories of age, gender, and skill, and they did not try to use the law to remedy fundamental social inequities (a temptation some American reformers found harder to resist).

While NRA debates over wage rates centered on whether worth and need were different for black workers than for white ones, the British system reflected a far more complex set of ideas about age, gender, and skill. It also encompassed the standard of an imagined "normal" worker against which mental and physical handicaps were weighed. The success of the Trade Boards in creating minimums and getting employers to agree to abide by them suggests that the relative homogeneity of British society made it possible for agreements to be reached on such norms.

The lowest wages paid under the Trade Board system went to children. Unlike the architects of the NRA, British reformers made no effort to use the act to eliminate child labor (the Factory Acts had already set minimum ages). Wage scales set general wage levels for girls and boys, to be increased gradually until they reached the age of eighteen. Writers of the wage scales assumed a normal life course for working-class girls involved entering the workforce before the age of sixteen. Children were assumed to work gradually through various "learner's" jobs until they either left the workforce or reached relatively skilled, well-paying jobs. In at least some cases, this system offered an incentive for employers to hire very young workers. Teenagers, on the other hand, had incentive to lie about their age: either to get employment (by saying they were younger than they really were) or to get higher wages (by claiming to be older).[27] Because of these normative assumptions, workers entering the workforce "late" (after the age of sixteen) presented a perplexing problem. Up until even 1937, the board made special provisions for "late entrants" who entered as learners after age sixteen. These women received the lower beginner's wage. Otherwise, board members worried that employers would not take them on at all.[28]

Pay scales also legitimized gender differentials and the conventions of sexual division of labor. Most Trade Boards assumed men should earn more than women. Sells stated that the general rule of thumb was that a man had to earn enough to support a family, while a woman needed only enough to live alone.[29]

In addition to supporting age and gender conventions, wage scales reified a hierarchy of jobs within the laundry by adopting conventions about skill. The arguments over what *skill* meant illustrate the often arbitrary nature of its definition. In the initial struggle over applying the acts to calender workers, laundry owners argued that calendering was skilled work. Miss Collet, of the Ministry of Labour, pointed out that the laundry owners had always previously claimed that calendering did not require skill. She made a further distinction, suggesting that calendering required "no skill" but rather "continuous carefulness."[30]

Both contemporaries and historians have condemned the Trade Board rates as being unjustly low.[31] However, making a fair assessment of this claim is difficult. Certainly the twenty-eight shillings an hour initially set by the board seems like a pittance. But for the lowest-paid workers, including those at the Millbay Laundry, it effectively doubled wages. Reformers hoped that minimums would provide a safety net for workers and that effective organization would give workers the power to negotiate wages well above that minimum. In the laundry industry, this strategy was only moderately successful. Union membership soared in the 1920s but then collapsed with the onset of the depression. Furthermore, the politics of local unions sometimes undermined the intentions of the act. At Millbay, Roberts agreed to the local union's insistence that married women be fired if they had husbands to support them so that unmarried women could have their jobs, a policy that effectively eliminated many of the most experienced, highest-paid workers.[32]

The NRA

Like Britain's Trade Board Acts, America's NRA was built on prior experience. Before the creation of the NRA boards, the Trade Board system had been most directly translated in the United States through various state-level efforts to establish legal minimum wages for women.[33] Most of the seventeen states that experimented with minimum wage laws before the implementation of the Fair Labor Standards Act adopted elements of the British system, including trade boards that set industry-by-industry wage rates. Laundries were among the industries singled out for inclusion in a number of states such as California, Ohio, and New Jersey.[34] Previous familiarity with such state-level systems contributed to the greater success of the NRA in these states.

By 1933, semiprivate advisory boards had become a familiar part of policymaking for many Americans. A number of laundry owners who wrote to the NRA drew analogies between the code boards and the War Industries Board that they remembered from World War I. Some laundry owners in the North were also familiar with various state-level labor advisory boards.

While the actual creation of the NRA had a certain improvised character, many of its mechanisms were borrowed from programs familiar to New Dealers. People who had a hand in giving shape to the NRA also had previous experience with either the minimum wage boards or with the War Industries Board.[35] Frances Perkins, Roosevelt's appointee for Secretary of Labor, had been deeply involved in the efforts to establish minimum wage boards in New York State, both as a member of the Women's Trade Union League and as industrial commissioner for the State of New York during Roosevelt's governorship. After the Second World War, Perkins continued to insist that the across-the-board minimum wage established by the Fair Labor Standards Act had been a mistake, that the United States would have been better off with a trade board system.[36]

Other women reformers who came to advise in the creation of the New Deal had also had experience with minimum wage boards in various Northern states.[37] In August 1933, Rose Schneiderman, who would soon be called to serve on the NRA Labor Advisory Board, took her seat on the New York Minimum Wage Laundry Board. New Yorkers had replicated the British Trade Board model down to its inclusion of "one lady distinguished for good works"–in this case Mrs. Daniel O'Day, "clubwoman."[38] Hugh Johnson, the director of the NRA, had gained experience with the board system as the War Department's representative to the War Industries Board during World War I.[39]

Although the NRA and the Trade Boards shared some structural and strategic similarities, they had different (though not antithetical) goals and were created within very different historical contexts. Advocates of the 1919 Trade Board Act extension had portrayed it as a response to long-term social and economic problems of British society, notably the potential of endemic poverty to fuel class warfare. It had also been motivated by a surprisingly wide consensus that British society should be made more just and humane—an appropriate "home for heroes," as the rhetoric of the time put it. Although many New Dealers (both policymakers and constituents) harbored similar hopes, the NRA was introduced as a temporary measure in response to an exceptional crisis in American capitalism. Its stated rationale rested upon the premise that it was a response to a "national emergency productive of widespread unemployment and disorganization of industry."[40] The economic rationalizations given for the NIRA suggested that the labor provisions would even benefit employers as well as the working class. Trade Board advocates had made no such explicit promises.

The administrators of the NRA always intended to include service industries under the code. To the NRA, the laundry industry seemed like an ideal subject for organization under a code. The industry employed almost 250,000 workers, and employers were well organized (into both national and regional trade associations), in contrast with many other service industries. Whether or not NRA officials believed their own rhetoric, they often told laundry owners that the industry was being used as a model for the creation of service codes. One cynical laundryman had a slightly different explanation; the laundry industry, he said, was "too big to hide."[41]

The NRA also counted on the desperation of laundry owners. The depression had had a devastating effect on the industry. A general decline in business affected everyone, but worst hit were the laundries that catered to the family wash. During the 1920s, this was the area that laundry owners had counted on to expand. Many of these operators had taken on large debts in order to invest in new machinery to offer cheap, technologically intensive services such as bagwash and rough dry work. Now impoverished customers saved a few dollars by doing the wash themselves, while better-off customers bought washing machines or gave their housekeeping money either to washerwomen whose services could be had for next to nothing or friends and neighbors who had fallen back on taking in washing to supplement their family income. As in many other industries, the situation was worsened by "cutthroat competition"–selling at or below cost to drive competitors out of business. Despite the efforts of the LNA, the practice spread rapidly during the 1920s. By the early years of the Great Depression it was epidemic, particularly in urban areas.[42]

After the initial hoopla of the president's announcement of the NRA, officials pushed trade associations to quickly develop industry-specific codes. The leadership of the LNA made a conscious decision to postpone any commitment until they could see how the codes worked in other industries.[43] They also asked business owners not to sign a so-called blanket agreement, in which employers promised forty cents an hour to their "industrial" employees. Afraid that such a stance would be condemned by the public, the LNA then quickly drew up and submitted a temporary code. However, the president's Re-employment Agreement for the laundry industry had far more acceptable terms than the blanket agreement–the lowest minimums in the Re-employment Agreement were fourteen cents an hour. Public enthusiasm for the NRA was running high, and many laundry owners signed onto a temporary agreement late in the summer of 1933.[44]

Unlike the Trade Boards, where interested parties brought their arguments and negotiated an arrangement, the process of writing a laundry code in the United States began with the trade associations. Research and writing of a draft code commenced in the spring of 1933 at the Laundryowners' National Association headquarters in Joliet, Illinois. The LNA was to be counseled on the creation of a code by three advisory groups: representing businesses, workers, and consumers.

The LNA proposed that, rather than a single set of national standards (as had been done in other industries), the country should be divided up into six regions and a distinction should be made between metropolitan areas and smaller towns. One of the consequences of this system was substantially lower wage rates in the South. There, the proposed fourteen-cent-an-hour minimum was the lowest wage proposed by any of the code boards.[45]

LNA lawyers initially argued that such unique arrangements were necessary because competition in the industry was regional, and the cost of living was lower in the South. Therefore, Southern consumers could not afford higher laundry costs, and Southern workers did not need higher wages to subsist. Such an argument was not substantially different than the logic that had given rise to lower Trade Board rates for Northern Scotland and Cornwall. However, the American subtext was all about race. Many Southern laundry owners were unwilling to pay their mostly African American workforce more than the fourteen cents an hour in the proposed code. They also had a tenuous foothold in a region in which custom and the racial dynamics of the Southern labor market made African American washerwomen a serious form of competition.

Many Southern laundry owners argued, however, that the rates were still not nearly low enough; that minimum wages even as low as fourteen cents an hour would mean that laundry owners could not compete with washerwomen and housewives. For those in Washington, D.C., the LNA's racial agenda was reinforced by a flurry of very explicit letters from Southern laundrymen. One laundry owner wrote to Hugh Johnson: "You probably do not know, but to make a profit in a laundry, with SLOW POCKY [sic] colored labor that we have to imploy [sic] means that we have a hard job to make any profit at all."[46]

Once the LNA had drawn up a proposed code, the NRA specified that it should be circulated among interested parties and published in newspapers for the benefit of the general public. Goodwill between the LNA and labor advocates began to evaporate as documents from the LNA's Illinois headquarters appeared. Alerted to the LNA's race-based strategies, disparate groups who believed that the NRA should be a tool for social justice as well as economic recovery began to mobilize. At the public hearing on the laundry code scheduled for November, they planned to expose the LNA's racist strategy to public scrutiny and convince the Department of Labor and the Labor Advisory Board to make raising the Southern minimum a priority. Clark Foreman, "Advisor on the Economic Status of the Negro" for the Department of the Interior, and his assistant and eventual successor, Charles Weaver, provided the impetus behind these efforts. Foreman and Weaver wrote a series of letters to Frances Perkins alerting her to the LNA's intention.[47] Uncertain of her commitment, they also went other routes. Through Foreman and Weaver's solicitation, black women's clubs, teacher's unions, and local YWCAs all sent letters and petitions to the code board.[48] They found Rose Schneiderman to be their most willing spokesperson. A Labor Department employee, M. J. Byrd, prepared for her use an extensive memorandum providing statistics and arguments about the status of African American laundry workers in the South.[49]

As the date for the hearing approached, the NRA also assembled a list of the organizations and individuals who were to represent workers, employers, and consumers. The final list reflected the sometimes conflicting priorities of the NRA, as well as its race, class, and gender politics. While finding an organization to represent the laundry industry was relatively simple, the categories of workers and consumers were more problematic. The AFL laundry unions, with their long, dismal history of relations with women workers, could hardly be said to represent the female majority in the industry. In the end, the AFL and the Teamsters' unions spoke for male workers, while Rose Schneiderman (as the only women on the labor advisory committee) represented the women. Although Schneiderman made it her business to speak directly to the issue of race, no provision had been made to let African American laundresses speak in defense of their own interests (the committee did hear from plenty of other workers–almost all of them Northern male drivers).

Issues of gender and class also governed the choice of representatives for consumers. The American Association of University Women was chosen to speak for consumers, suggesting that the administrators thought consumption to be the business of middle-class white women.[50] The inclusion of consumers represented a significant departure from previous uses of the Trade Board model. Their presence could be interpreted in a number of ways. While consumers' input had very little direct effect on the eventual shape of the code, their inclusion likely had a political motive. While the British Trade Boards were intentionally designed as an encounter between representatives of two different social classes, including consumers in American boards helped diffuse class tensions, carrying the message that the NRA (and the New Deal) transcended class boundaries.

On November 21, 1933, the various interested parties met in Washington, D.C., for the code hearing. Unlike the private negotiations of the Trade Boards, the laundry hearing was a public event (evidence perhaps of the American tendency to confuse democracy with public spectacle). Only half the transcript survives, but its more than five hundred pages are enough to suggest the peculiar qualities of that hearing. Many of the participants had little idea of what they were there to accomplish or how to obtain their ends. The code administrator apparently treated the occasion as a forum in which all interested parties could feel that they had participated. Unlike the Trade Board meetings, which forced participants to negotiate at the table, the precirculated LNA plan restricted the discussion to attacks and justifications of the plan punctuated by statements by people who had not bothered to read it or did not even know it existed.

Amidst all the varied statements, the question of Southern minimums emerged as a central point of conflict between labor interests and laundry owners. Southern laundry owners made little effort to conceal their arguments. They and several representatives of the NRA took the stand to testify that African American workers were less productive than white workers and that the unregulated washerwomen competing against Southern laundries were willing to work for meals or pennies a day.[51] They suggested that if the point of the NRA was to regulate competition, such regulation should also extend to those that took in washing.

Representatives of reform interests protested the Southern arguments. A Women's Bureau representative, speaking for Mary Anderson, its head, pointed out that the Southern scheme undermined the essential point of wage minimums, since such low wages could not increase the purchasing power of workers. Rose Schneiderman followed with a characteristically blunt statement. Like British reformers, she used the rhetoric of citizenship and social responsibility to put public pressure on laundry owners. She began her testimony on a note of sarcasm: "I almost wept when the professor last night outlined the dire grief and poverty of the laundry industry in the South." She then pointed out that the laundry industry had paid extremely low wages long before the depression and that many employers across America were also in debt but were willing to raise wages for the good of all. Taking a shot at laundry owners' arguments about free enterprise, she suggested that since the proposed wages were clearly too low to live on and the government would end up making up the difference in relief, the Southern laundrymen were essentially asking the government to subsidize their payrolls. She completed her testimony by expressing her prayer that the "gentlemen" of the LNA would recognize their responsibility to "bring to these thousands of women working in the industry some hope."[52]

As if to confirm that the Southern laundrymen's stance was not a renegade diversion from the LNA's policy, in the aftermath of the hearing a lawyer from that organization sent the NRA a brief summarizing its basis for wage rates. The brief pulled no punches in laying out the racial basis for the minimum: "Approximately 95% of the productive labor in the laundries in the south are negro. This is a class of inefficient labor, entirely unskilled and slow." The brief went on to say that Southern employers would prefer to hire white workers who were more productive, but "there is a social distinction making the employment of white labor impossible in departments where colored employees are employed."[53]

Hope was not, in fact, forthcoming for many laundry workers (or laundry owners). While the laundry code was being negotiated, cases challenging the constitutionality of NRA service codes began to appear in the courts. Those most closely watched by the laundry community involved the setting of price minimums by the dyeing and dry-cleaning code. Anxious to avoid providing an opportunity for an all-out constitutional challenge, the NRA revised its proposed laundry code to include what came to be known as the "local option." The local option stipulated that instead of the LNA signing a code that would then apply to all laundry owners, local groups would have to obtain an agreement from 70 percent of the laundry owners in their area that a "state of emergency" existed and that the code should be put in place.[54] Code officials were circumspect about the constitutional reasons for the local option. Instead, they offered the nearly meaningless rationale that the "code could be made to operate better in a local area."[55]

The peculiar characteristics of the industry made the local option palatable to the LNA. It was willing to forgo a basic national standard of competition–the very provision of NRA codes that made them attractive to large manufacturers–because competition in the laundry industry was always local. Southern laundry owners who opposed the fourteen-cent minimum most obviously benefitted. They, in turn, were willing to forgo the goodwill of a minority of white Southern workers who argued, along with Mrs. M. A. Focht, president of the Birmingham Laundry Workers' Union, that the government was obligated to provide "white skilled or productive labor" with a "decent living wage," even if it did not do so for African Americans.[56]

Panic on the part of the NRA briefly subsided when a New York judge returned a ruling on the most closely watched dry-cleaning case in April 1934, upholding an injunction against price fixing. It also turned out that the case could not be used to challenge the authority of the federal government to regulate intrastate commerce, because the laundry in question had plants in both New York and New Jersey. Hugh Johnson announced to the industry that "uncertainty as to enforcement by legal proceedings must end."[57] The NRA's relief was short-lived, however. Additional constitutional anxiety led to an executive order directing the NRA to stop enforcing price controls in service industries. Soon afterwards, the issuance of the Darrow Report, severely critiquing the NRA, led to rumors that the service codes had been withdrawn completely.[58] In June, the NRA announced "The Code Still Lives," but the LNA was hard put to find signs of life. Curiously, signatories of the President's Re-employment Agreement who had never voted for the local option continued to be responsible for subscribing to the conditions of that agreement. The trade journals distinguished between the Re-employment Agreement and the NRA as the "old eagle" and the "new eagle."

For all intents and purposes, the service codes were dead nearly a year before the Supreme Court's decision on *Schecter v. United States* would put the rest of the NRA out of its misery. The NRA nevertheless began to push for state-level legislation that might accomplish some of the aims of the national program. Because the laundry industry was exempted from the minimum wage established by the Fair Labor Standards Act until 1966, these state laws would be the only wage legislation for laundry workers over the next thirty years.[59]

Surprisingly, in some states laundry owners themselves pushed for minimum wage legislation. In the course of this brief "experiment," the LNA had also come to a curious reversal of position on the virtues and drawbacks of the code. In the beginning, laundry owners had vigorously opposed the wages and hours stipulations of the code while touting the price minimums as their reward for compliance. By the middle of 1934, though, they were decrying state-mandated price fixing while advocating minimum wage laws. As the NRA crumpled, local trade associations in Northern and Midwestern states began petitioning state legislatures for minimum wage legislation or for the enforcement of legislation already on the books. Many laundry owners had come to realize that minimum wages were a more effective means of driving "chisellers" out of business than price minimums.[60] Meanwhile, as would be expected, race continued to play a central role in Southern states' rejection of minimum wage legislation.

In 1935, as the collapse of the NRA seemed imminent, Howard Myers of the NRA's Labor Advisory Board approached Mary Anderson, director of the Women's Bureau, about carrying out a study on the "adequacy" of the NRA minimum wage in the laundry industry. Although Anderson was initially hesitant, she agreed after it became clear that the impetus came not from apologists for the NRA but rather from the civil rights and reform communities.[61]

It was apparently Myers's intention to better prepare the Labor Department to construct successor programs to the NRA. The resulting report gave the women's reform and civil rights communities a small form of revenge against the machinations of the LNA and the Southern laundry owners. Anderson corresponded politely with the LNA while her staff collected damning evidence about wages and profits in Southern laundries.[62] The accumulated data showed that while Southern wages were the lowest in the country, Southern consumers did not reap the benefits–the difference between wages and profits had been going into the pockets of Southern laundry owners.[63]

✕

State intervention became more and more a part of the laundry business during the interwar years. Although workers might have seemed to be the main beneficiaries of these measures, many laundry owners saw regulation as a means to create controlled competition in an erratic business climate. They welcomed such regulation if it could be applied evenly across the industry.

Although similar strategies were used in Britain and the United States to gain employer cooperation, British efforts met with more success. As in previous periods, constitutional structures between the two countries were an important cause of different outcomes. Even the conditions of the Great Depression were not enough to convince American judges that laundries were a form of interstate commerce. But the politics of race in the United States also provided a wild card that helped guarantee the NRA's failure to create a fair and effective plan for the laundry industry.

Compliance and Enforcement at the Other End of the Stick

DESPITE NRA AND TRADE BOARD RHETORIC ABOUT DEMOCRATIZING policymaking, most members of the industry had little or no input on the legislative end of the process. Their relationship with the state and their understanding of the law resulted from their experience at the other end of the stick: It was shaped by reading about policies in newspapers and trade journals, by receiving Trade Board orders in the mail or going to the post office to pick up a blue eagle emblem, and by meeting inspectors and other officials. Although implementation and enforcement carried the threat of coercion, employers and workers were far from passive subjects of policymakers' intentions. Together, employers, workers, inspectors, and judges remade policies through interpretation and interaction.

The NRA and members of the Ministry of Labour understood that a well-structured program was only the beginning of effective social policy. The long-term success of both the NRA and the Trade Boards depended on gaining the cooperation of those affected by the laws. The British government was able to achieve reasonably good compliance with the Laundry Trade Board. In contrast, the NRA rapidly lost the cooperation of even its more enthusiastic supporters. Laundry owners made the immediate difference between the two efforts clear: the Trade Board could guarantee enforcement while the NRA could not.[1]

The NRA is admittedly anomalous, even in American history, for the rapidity and drama of its failure. However, the particulars of that unraveling reveal more systemic and enduring differences between the ways that Americans and the British understood the nature of the law and their relationship with the state. Moreover, neither the NRA nor the Laundry Trade Board was created in a historical vacuum. Whether disputes involved a Texas laundryman tired of local corruption or a formerly socialist London manageress, all the participants drew upon their previous experiences and beliefs when making decisions about how to act in regard to these programs. Likewise, the successes and failures of both programs had lasting impacts beyond their immediate purview.

Inherent assumptions built into the laws complicated implementation and enforcement. The NRA first designed the program's overall blueprint with monopolistic, vertically integrated industries like cotton textiles and automobile manufacturing in mind. Enforcing job-specific wage standards was relatively easy in workplaces in which rigid divisions of labor were in place or hundreds of employees did the same job. It became far more problematic in small-scale factories (like laundries) without rigidly standardized practices, in which employees sometimes did more than one job and mechanization, with its attached connotations of skill, proceeded in a highly idiosyncratic and piecemeal fashion.[2] Although Parliament and the Ministry of Labour wrote the Trade Board Acts with "sweated industries" in mind, the nonproductive nature of laundries and their complex labor organization caused similar problems in Britain.[3]

Whatever might be said in trade literature and government publicity about regulation's eventual benefits to the laundry industry, the ordinary laundry owner and manager found new regulations a nuisance, not only costly but also requiring careful recordkeeping and organization. For instance, the Trade Board system demanded business owners compile careful payroll records–something many laundry employers had never done.

To accommodate the complexities of the laundry trade, both the NRA and the Trade Boards utilized multiple categories and schedules. Particularly in the United States, administrators left laundry owners to figure out how the idiosyncracies of their business fit into categories created by these laws.[4] New laws also reached into laundry owners' pockets in more unexpected ways. The LNA, for example, assessed its membership a fee to pay for the research used in drafting the code.[5]

Moreover, many employers and workers had only the vaguest idea of how government and the law functioned in general, so they were in a poor position to interpret how new regulations applied to their particular situations. As a group, laundry owners and managers were generally not well educated. Many had worked their way up from being inside workers or drivers. Until the growing importance of regulation, their success had depended only on specific knowledge of laundry processes, not on a mastery of the principles of labor law or business theory. And many were simply lazy, stupid, and stubborn. For men such as the British employer who paid his fifty workers for forty-eight weekly hours of labor but never cared to know whether they worked ten hours or eighty, regulation added another bothersome task to their businesses that many were simply unwilling to master.[6] Others, desperate to balance the books, resorted to rationalization or elaborate subterfuge. They compared the threat of prosecution for labor violations with the more immediate fear of snarling creditors and made their choices.

Ignorance of and resistance to the law created a twofold challenge for government agents. The success of state regulation depended on the proper use of many tools: education in the particulars of the law, measures to obtain widespread voluntary cooperation, and mechanisms of coercion and punishment when all else failed. In the best-case scenario, employers would understand the law and would adopt its practices voluntarily. It was in the state's interest to minimize the number of employers actually taken to court. Too many prosecutions would engender ill will and cost the government time and money. Moreover, a frequently conservative British judiciary, hostile to labor reforms, often ruled in favor of defendants.[7]

In the United States, NRA officials also tried to avoid the courts. Fear that a prosecution might turn into a constitutional battle created incentive to avoid outright coercion. Instead, administrators looked for ways to convince employers they should sacrifice immediate profits for what was portrayed as the greater good of the society and their own long-term gain. They also used routine inspections as an educational tool, issuing a warning rather than a summons for first violations.

By bringing trade associations and unions into the policymaking process, the NRA could use these organizations' mechanisms for disseminating information to publicize new laws. The editors of trade journals were asked to print statutes for their readers. Writers for such publications also took it upon themselves to explain the workings of the law, or they asked fellow laundry owners or, more rarely, government representatives to contribute articles.[8]

Using trade journals for educational purposes could, however, be a dangerous strategy for the government. British editors were more cooperative than their American counterparts in publishing articles to support the government's aims (or at least to tell laundry owners that they should obey the law). The editorial stance of the American journals varied widely. The opinions expressed towards the NRA in *Laundry Age,* the most widely read journal, varied from glowing enthusiasm to accusations of communist influence in the New Deal. American editors also reprinted clippings from other publications and letters from readers to voice particularly strong antigovernment sentiments–giving the appearance that resistance was not centered in the LNA but rather a generalized sentiment among laundry owners.[9]

Government officials found it most difficult to reach workers with information about new laws. In Britain, the Trade Board Acts specified that rates be posted in a conspicuous place, but evasive employers used a variety of strategies and excuses to get around the law. A favorite scam was to tell the inspector that the notices had been hung but steam had peeled them off the wall or that the laundry was in the process of being whitewashed (for sanitary reasons, of course), necessitating the removal of all placards. One particularly devious laundryman could rightly claim that he had posted the notices–the inspector found the men's notice in the women's cloak room.[10]

The NRA Laundry Code had a similar provision, which specified that "every member of the trade shall display a placard adjacent to the time clock, or employees' entrance and at three other prominent locations in his place of business, stating the minimum wages and maximum hours of labor which govern the operations of said member."[11] Unions also reprinted the labor provisions, using them as an organizing tool. AFL pamphlets trumpeted "Uncle Sam wants you to join a union!"[12]

To overcome employer evasiveness, both the NRA and the Trade Boards mounted more general campaigns to convince employers that complying voluntarily would serve the greater good of society. The American version might best be described as an appeal to patriotism (urging owners to cooperate out of loyalty to the state). Laundry owners who wrote to the NRA often voiced their motivation in terms of their duty as citizens to support the goals of the state. Many echoed the Rhode Island laundryman who wrote directly to the NRA's head, Hugh Johnson, announcing his willingness to comply with the code despite lack of enforcement because laundry owners "are backing up President Roosevelt, Hugh Johnson, and the NRA 100%."[13] The image of cooperation with the NRA as a patriotic act was fostered by a massive publicity campaign, which not only introduced the program's goals to the public but also used displays like parades and symbols like the blue eagle to cultivate and exhibit general support.

The NRA also gave the public a language with which to talk about cooperation (using slogans such as "We do our part") and to denigrate noncooperators (calling them "chisellers," or "cutthroat competitors"). Unlike similar wartime programs, New Dealers spiked the "common good" theme with an explicit economic incentive. Over and over they argued that improving the economic well-being of workers would, in the long run, improve employers' economic status.

The patriotism strategy attempted to minimize differences in interest, asserting the possibility of the government determining a common good with advice from its citizens. Many members of the business community were, however, unwilling to commit absolutely to these ideas. The patriotism argument ran up against the frequently made suggestion that regulation was contrary to the belief in the "American system of private initiative and unlimited opportunity." One particularly vehement editorial vividly evoked the specter of communism. The author told the readership of *Laundry Age* that the "better type of laundryowner, who is overwhelmingly in the majority" resents criticism from "professional social workers, union labor organizers, and political 'pinks' praying with their faces toward Moscow."[14]

As the NRA came apart, the gap between public patriotism and private self-interest became more readily apparent to laundry owners who had complied with the code. They derided not only "chisellers" in the industry but also consumers who acted out of self-interest. The president of the Pennsylvania Laundryowners' Association described the behavior of the latter: they "will attend patriotic meetings, openly announce their loyalty to the Government, shout loudly 'we do our part,' tell the neighbor over the garden fence that they uphold the NRA but will give their laundry to a chiseller to save five cents on the whole bundle."[15]

C. T. Thomason, a Texas laundryman who shared a long-running exchange of letters with NRA officials in Washington, made the point more bluntly. Houston laundrymen, he said, had concentrated on fixing prices rather than on raising wages. The result was that workers were no better off than before. Moreover, high prices encouraged customers to avoid laundries altogether. Thomason said they instead "get a woman to do it on the yard." As evidence, he offered a three-way conversation he had had with a customer and "one of the NRA compliance women workers" who failed to practice what she was employed to preach. He asked the compliance worker where she sent her laundry, without revealing his occupation: "Then she showed her true color. This is what she said. 'I send my laundry to a negro washwoman. She do it for half the price.'" Thomason asked her if this was fair to the NRA and she replied: "I do not care. Nobody are [sic] going to tell me what to do." Thomason then revealed that he was a laundryman, "and how quick did she leave. And she is not the only one I have met."[16]

The idea of public displays of patriotism was not alien to British political culture, but little use was made of it in convincing laundry owners to cooperate with the Trade Boards. Instead inspectors, representatives of the Ministry of Labour, and trade journals tied the idea of the common good to a subtle (or perhaps just poorly defined) set of ideas about Englishness. They argued that while class interests might separate employers and employees, all shared a common culture. The obligations to look after the welfare of the less fortunate and to obey the law were two salient features of that culture.

This formulation was reinforced by the (somewhat paradoxical) threat that those who did not practice responsible capitalism could lose social standing while those that did could enhance their position in society. Over and over employers of the "better sort" were contrasted with "marginal employers." In preparing testimony before Parliament for the inclusion of laundries under the Trade Board Acts, factory inspector Clara Collet combined the status argument with a then very current allusion to animal abuse: "The policy of paying young women semi-starvation wages is unwise from the employer's point of view (unwise as driving a van with a half-starved horse)."[17]

The idea of an English polity also led inspectors to consider some groups to be outside their realm of responsibility. Most conspicuously, Trade Board inspectors did not inspect Chinese laundries (although at least some of them did appear on inspectors' lists). For many inspectors, the Chinese did not belong to the political community with which the law was an agreement, and they were generally thought to be incapable of understanding or agreeing to the intentions of the acts. Because Chinese laundries competed with white-owned laundries and sometimes employed white ("English") women, the government would, however, authorize an investigation if it was specifically requested by an employee or competitor.

In 1937, the Ministry considered prosecuting Mr. Fong Wah of Newcastle-upon-Tyne after a complaint was issued "which revealed heavy underpayment not only to the complainant but also to an English girl." The inspectors debated the complainant's viability as a witness, and they decided he would probably be adequate since he was "not a Chinaman" but rather an "Anglo-Chinese halfcast."[18] The notes of the lady inspector assigned to the case suggest that she was reticent about getting involved because the situation involved cultural elements beyond her expertise. The complaint itself was partly an act of revenge for an unpaid gambling debt. The inspector also held out little hope of reeducating Wah to run his business along Trade Board lines. She described him as "possessed of native cunning" but otherwise "illiterate"—neither speaking nor writing English.[19]

Administration of the acts offered another arena in which the meaning of waged work could be negotiated between individual employers and the state. The law contained a specific clause allowing employers to claim exemption from paying full rates to physically or mentally handicapped employees or those who were otherwise deemed, in the language of the inspectors, "sub-normal." Although official policy required a doctor's certificate attesting to physical or mental handicaps that might lower a worker's productivity, inspectors were actually empowered to grant exemptions on their own. When reporting on workplaces, they made determinations on whether workers were "worth" the posted trade board rates.

In one example of discretionary practice, inspectors were unwilling to seek prosecution of an employer who had underpaid one of his washmen after investigation revealed that the worker, who had been employed by the laundry for almost fifteen years, was nearly blind. He found his way around the washhouse by touch and long familiarity, but he had caused his employer some difficulties with customers by mixing up white and colored goods in the same washes.[20] Other employers were also granted exemption after sending their workers for interviews with inspectors. After an investigation of a laundry in a Cumberland coal town revealed a number of violations, inspectors considered making a blanket exception for the locality because "with the exception of one worker, all workers were sub-ordinary owing to age, lack of experience, poor physique, or lack of diligence."[21]

This interpretation of the law stemmed from the British spirit of paternalism. Inspectors, employers, and community members all assumed that business owners would take in marginal members of the community for reduced wages rather than let them starve. Such practices also had a practical basis in a society full of disabled returning veterans and a working class warped and stunted by the hazards of industrial poverty. The racial homogeneity of British society also made it a workable strategy because employers could apply the idea of "subnormality" on an individual basis rather than to an entire race.

Where voluntary cooperation was lacking and terms for exemption did not apply, the threat of coercion was necessary to maintain the authority of the state. Both the NRA and Trade Board administrators understood the necessity of enforcement, but they appreciated that the rules contained in statutes had to be interpreted in consonance with community understandings of fair and necessary application. The arguments used to gain voluntary cooperation also shaped and limited the coercive means representatives of the state could use.

By all measures, the NRA failed utterly in enforcing compliance. Problems of disorganization were far worse in service industries than in manufactures.[22] In the name of efficiency and in order to cater to the semivoluntaristic strains of American political culture, industries were expected to organize themselves into local compliance boards before the government could attempt enforcement. In the beginning, NRA officials had promised stiff fines against violators and prompt responses to complaints. But in reality, no area of the country ever got a fully functional compliance system for the laundry industry under the federal NRA plan (although the New Jersey State NRA board for laundries was by all accounts very efficient in enforcement).[23] In desperation, Philadelphia laundry owners pooled their money and hired an inspector at sixty dollars a week. Even then, they found that the local compliance board failed to levy a single fine.[24] The government was very cautious about enforcement because of the tenuous legal standing of NRA codes for service industries.

The debacle of the NRA's enforcement policy was a consequence of the problematic history of uneven labor law enforcement in the United States. Even in Northern states with long histories of factory legislation, effective application was often sporadic and irregular. Factory inspectors sometimes got their jobs through patronage, not diligence or skill, and so the fortunes of state labor departments waxed and waned depending on the politics of incumbents. The depression aggravated these problems by decreasing funds available for inspection and prosecution.[25] In the South, all such problems were magnified by greater poverty and more explicit racism. This contrasted with the highly professionalized British inspectorate, whose fortunes remained relatively immune to political change.

Nevertheless, individual laundry owners often held more hope for federal enforcement than the corrupt local systems they were familiar with. In another one of his letters, C. T. Thomason emphasized the patterns of corruption in local politics. Believing local code authorities to be corrupt and in league with the owners of the larger Houston laundries and the Chamber of Commerce, Thomason wrote directly to

Washington D.C. He pleaded that the Houston area be provided with a local code administrator who was not a member of the Chamber of Commerce and who was "honest in character and fair to employers and employees alike." In addition, the NRA should send an "under cover man and woman" to Houston to investigate. He suggested that the investigators interview workers in their homes because he had been told that "operators and clerks have been threatened that if they reveal the truth about working conditions they will be expelled and black balled."[26]

American trade journals questioned the ability of even the federal government to enforce its own laws. They offered up the case of Prohibition–the inability of the government to enforce effectively the Volstead Act–as an example all too familiar to many Americans. "You don't want another Prohibition fiasco," was the way one article put it.[27]

As in Britain, the politics of enforcement in America included the issue of discretionary exemption. Surprisingly, laundry owners do not seem to have been vehemently insistent that Chinese laundries be included under the NIRA. In assessing the situation in Chicago, Women's Bureau investigators found that many white laundry owners believed that "Chinese laundries were gradually fading out as younger Chinese are not interested in doing laundry work."[28] In cities like San Francisco, NRA officials did try to get Chinese laundry owners (and the tongs that often represented them) to subscribe to the act, though with limited success. James M. Hanley, lawyer for the Tung Hing Tong of San Francisco, telegraphed Hugh Johnson in the summer of 1933 to say that his clients wanted a separate code appropriate to the "Chinese way of doing business." Hanley suggested that the tong's willingness to cooperate was mostly a gesture of goodwill since "there is no unemployment among skilled Chinese laundry workers in San Francisco."[29]

Instead of focusing on Asian laundries, many laundry owners wanted some form of regulation for washerwomen and domestic servants. In one sense, the argument was not so far-fetched. The government could have required anyone taking in laundry to have a license. Southern laundrymen also pointed to competition from African American washerwomen. The same correspondent who had complained to Hugh Johnson about his African American workers being slow and lazy now also supported government regulation of washerwomen. He thought it a fine idea but opined that enforcement might be a problem, writing that (in an image that conjured up military reconstruction) "it would take a small army to see that they obey orders."[30]

The British were far more effective in using the threat of prosecution. They particularly recognized that public acts of defiance could not be tolerated. In choosing whether to prosecute, they gave serious consideration to the question of intent. Employers who had violated the Trade Board Acts out of ignorance or because they were in desperate financial circumstances seldom found themselves in court. Employers who openly defied the law or the inspectorate, however, were most likely to face a judge.

The records of cases investigated for possible prosecution in Britain illuminate the delicate boundaries of enforcement and the inherent problems of discretionary authority. They also reveal how identities such as race, class, and gender could condition the way lawbreakers saw themselves in relation to the state. Many violators recognized the authority of the state but thought they somehow deserved exemption.

One of the most complex and fascinating of the Trade Board cases illustrates both the importance that administrators and directors of the trade association put on public promises of enforcement and the ways in which individuals brought their own understanding of the law to their dealings with the state. In January 1921, an inspector paid a visit to the Randell House Laundry in London, where she spoke with the manager, a Mrs. Caroline Jameson, and examined the company's records. The inspector found that although all the workers were paid at or above the board minimums, the Trade Board notices had not been posted. She returned twice that summer and was denied admission by Mrs. Jameson on both occasions. Seeking to find out more, the inspector made inquiries around the neighborhood. She came away with the impression that Mrs. Jameson was a good employer who provided jobs for women who would not have been hired elsewhere, or, as Mrs. Jameson more colorfully put it: "I employ about 45 here. Amongst them the halt, the lame, and deaf."[31]

Further inspections revealed the truth of her claims. Her employees included several mentally retarded young people, a single mother of a young baby, and an ancient woman who had threatened suicide should she be let go. To the inspector's apparent embarrassment, Jameson had also provided a job for a woman who had worked as a secretary at the Ministry of Munitions and as a librarian at Montagu House until she had been pushed out of these jobs by returning male workers. The laundry was also found to be in poor financial shape, and it became clear that Mrs. Jameson was supplementing the receipts out of her own pocket to keep the laundry's board of directors from closing the business.[32]

It would seem that Mrs. Jameson was doing her best to ameliorate the conditions of workers–indeed, at her own expense–and that inspectors might excuse her minor eccentricities, as they had done for a number of other employers. Certainly in her own mind, Mrs. Jameson thought she was doing her part for the greater good. Instead, however, inspectors continued to visit her laundry and to push for prosecution. They did so because she willfully and publicly defied the authority of the acts, not only in the conduct of her business but also by writing inflammatory letters to newspapers, trade journals, and members of the government. Additional pressure to prosecute was applied to the inspectorate by the head of the Federation of Launderers, who personally wrote to the Minister of Labour, threatening that members of the federation would "ignore the Trade Board unless Mrs. Jameson is brought to book."[33]

Mrs. Jameson framed her defiance in the same language of Englishness and community that the government had used, but she questioned the ability of the state to know better than she how that good should be achieved. Like a lot of her American counterparts, she suggested that government control undermined or failed to reward the best aspects of individualism (though her version made no mention of Moscow or the American way of life, of course). In a letter to the Minister of Labour, she described her qualifications for drawing such a conclusion: "For forty years I have been in touch with the working class and am proud of it. At the same time I have had the practical experience which the extremely irresponsible Labour Leaders lack. I have been proud to call myself a 'Socialist' in the past and for some time belonged to the ILP in this District." She charged that the trade unions were motivated by selfishness and that the government and employers had no understanding of the plight of a predominately female workforce. In contrast, she was a "woman's woman," different from the male laundry owners who dominated the trade associations. "I was laughed at in scorn by many in the trade. Called a pamperer of women, a *slobbering nurse*," she wrote.[34]

Jameson tried a different tack with members of the trade, describing herself in an open letter to the *Laundry Journal* as a "free trader" standing with "a gun in each hand and a sword in my mouth." The editor was not appreciative of the masculine imagery or the note of open defiance, however. He published her missive along with a note telling her that the *Laundry Journal* was a "responsible paper" that could not "advise its readers to break the law and risk the consequences."[35]

Eventually the decision was made to prosecute Jameson. The magistrate (like many of his conservative brethren) sympathized with her position as a representative of capital, defying the incursions of a socialist government. He used the opportunity to give what one of the inspectors described as "a long-winded speech about the evils of the current incarnation of the Trade Boards." However, he also distanced himself from Jameson's particular form of defiance by describing her as a "conscientious but misguided person."[36]

In general, people seemed to treat Jameson as an eccentric, but they did not see her as harmless. Laundry owners' and government representatives' unfavorable attitudes towards her were probably exacerbated because of her gender. Both laundry owners and inspectors (most of whom were female) seemed at the same time angered and bemused that a woman should take such a stance. Laundry owners were unlikely to rise to Mrs. Jameson's defense because she was not "one of the boys." Inspectors appeared to generally deal more strictly with women who defied the acts than with men who broke the same rules—perhaps because it was assumed that women would more "naturally" understand and sympathize with the intentions of the acts. Jameson was not the only female manager who was treated more harshly than her male counterparts.[37]

Laws worked best in areas where employers were well socialized in the dominant political culture and had previous experience in dealing with effective state regulation. In Britain, inspectors had their greatest problems with small-town laundry owners and minor operators who did not belong to trade associations and were protected by a network of other locals who put the interests of their class ahead of the authority of the state. In 1928, the ministry tried unsuccessfully to prosecute the manager of a small laundry in the town of Lydney. Unlike Mrs. Jameson, Percy T. Arnold, part owner and manager, had intentionally underpaid employees, lied about the laundry's records and about posting the rates, and had tried to undercut the prices of another laundry in the neighborhood that was paying Trade Board rates. The inspector, Miss Pillers, noted that the laundry seemed to be in poor financial condition. Inspectors generally tried to avoid the public humiliation of a court trial when it involved employers who had underpaid workers because their businesses were failing. Prosecution was recommended in this case, however, because of Arnold's willful efforts at deception.[38]

A panel of three local magistrates heard the case. Miss Pillers, in the company of another female staff member up from London, was delegated to argue the government's side. In the audience sat members of the community, including representatives of the press (Trade Board prosecutions were rare enough to be news in small towns). In Arnold's defense, his financial partner, Mr. Orchard, gave a memorable testimonial to his character, explaining why he continued to admire Arnold despite losing an investment of nearly five hundred pounds and why the court should similarly excuse him. "He is a real white man," the *Lydney Observer* quoted Orchard as saying. "He has been a really hard-working man. On several occasions I have come to Lydney and found him sweating in his shirt and trousers trying to get the business on a firm footing." *The Citizen* also quoted Orchard's words but embellished upon them: "He is a white man from his hair to his toe nails." The reporter also cited testimony in which Arnold explained to the court that he had put his own money into the laundry and was on the brink of ruin.[39]

The magistrates refused to find Arnold guilty despite overwhelming evidence to the contrary. The inspector thought that the magistrates' lack of sympathy with the "labour view" was undoubtedly the reason. An inspector's later visit found Arnold still at the laundry–he had gone bankrupt and was now working in the washroom. The inspector noted that he was being paid less than trade board rates.

⌘

During the 1920s and 1930s, British laundrymen like Arnold found it more and more difficult to conduct their business beyond the reach of the state. As regulation became more extensive, various interested parties tried to assess its effects. In Great Britain, the impact of the Trade Boards seemed immediately apparent to many in the trade. Looking back on the development of the industry on the occasion of George V's silver jubilee, in 1935, J. J. Stark counted the Trade Board Acts among the most important events in the industry's history. He stated that regulation had had an impact in encouraging mechanization. He also believed, as did many other laundry owners, that regulation had pushed laundry owners to better organize their businesses.[40] The threat of a visit from the inspector encouraged managers and owners to keep careful records for the first time. The Federation of Launderers benefited greatly from the Trade Board system and grew rapidly in both size and power after 1919. Reformers had intended the Trade Boards to encourage unionization. While the acts did seem to have this effect initially, by the 1930s membership had dropped precipitously.

It would be difficult to draw similar conclusions about the aggregate impact of the NRA on the American laundry industry during the program's short-lived tenure. Certainly the NRA affected the lives of individual laundry owners and workers. Some laundry owners abided by the terms of the NRA, despite incurring heavy financial losses, out of a sense of principle. At least some of the workers in those laundries were spared unemployment. Neither chiseling nor compliance guaranteed economic survival in the topsy-turvy world of Depression-stricken America.

The long-term significance of the NRA was transitional. In general, like the New Deal, it opened the door to increased state regulation. Frances Perkins later wrote that she saw the primary contribution of the NRA as educational–it convinced people that this kind of regulation was possible.[41] Certainly this was borne out in laundry owners' later efforts to get state-level minimum wage laws. Most importantly, section 7A of the National Industrial Recovery Act, which guaranteed the rights of workers to organize, opened the door for the most extraordinary and successful drive for unionization in the history of the industry. With the passage of the Wagner Act, in 1935, the freshet that had sprung forth with section 7A became a flood.

10

Women and Men and Unions

IN 1937, ROSE SCHNEIDERMAN DECLARED VICTORY IN WHAT SHE would later describe as the New York Women's Trade Union League's "Thirty Years' War" to organize female laundry workers.[1] A hardened participant in the league's often futile unionization efforts, Schneiderman knew how much there was to celebrate: officials of Local 300, United Laundry Workers (CIO) claimed almost 27,000 members, a 2,000 percent increase from the 1,150 members the International Laundry Workers' Union (ILWU), its AFL predecessor, had claimed three years previous.[2]

While laundry drivers had had some successes in forming stable organizations before the 1930s, most inside workers' unions had long been disorganized and poorly funded.[3] During the late 1930s, however, New York's Local 300 was one of the largest and most dramatic examples of the thriving laundry unions that could be found in all corners of the United States, for the first time in the history of the industry. More remarkably (from the perspective of the labor movement), Local 300 was an interracial union with a female majority. Its earliest members had all worked in the poorest paid segment of the industry: towel and linen supply laundries.[4]

The New Deal was indisputably the immediate cause of this change. Passage of the National Industrial Recovery Act in 1933 set off an initial wave of organization because section 7A guaranteed workers the legal right to organize. After the end of the NRA, the passage of the Wagner Act (National Labor Relations Act) in 1935 and its confirmation by the

Supreme Court in 1937 gave unions and workers extensive new rights. Both these measures also sent the message that the state now supported organized workers–a sharp contrast to the antiunion atmosphere of the 1920s.

Many workers testified to the extraordinary change "the Union" brought to their work lives. Shorter hours, better wages, job stability, and protection from capricious or hostile employers were just some of the benefits. Despite their complaints, employers also profited. The Wagner Act had been passed in the wake of the intensive, nationwide wave of strikes in 1934 (strikes which included laundry workers). Its mechanisms were explicitly designed to stabilize labor relations. The language of the act provided arbitration in cases of "unfair practices," whether such practices were on the part of workers, unions, or employers. Knowledgeable employers could therefore use labor contracts and union officials to discipline workers. Instead of the individual bargaining that had been characteristic of the industry, negotiations were now put in the hands of shop stewards, professional labor unionists, lawyers, judges, and accountants. Arbitration provided a means of settling differences and also inserted union officials into the day-to-day business of running a laundry.

In well-organized cities like New York, the Wagner Act had a transforming effect on the industry. Owners of small laundries who had successfully sidestepped the NRA and various state-level attempts to regulate wages and hours now found themselves bound up in citywide contracts. A "joint board," consisting of representatives from groups of owners and managers, created these contracts. In effect, laundries were now subject to a private NRA, which enforced the contracts with the threat of both strikes and legal actions.

The new CIO unions also offered other potential benefits not available under even the best-administered labor laws. Laundry worker Emma Crosby was blunt in her appraisal: "The state minimum wage law did not protect women and particularly Negro women from insults from their employers. In laundries, a job and steady work depend on being 'friendly' with the foreman or even the employer himself." She argued that the only protection against what would now be called sexual harassment was a strong union.[5]

Although the interwar period in Britain was marked by dramatic labor activism in other industries, neither the Trade Board Acts nor broader social and political changes seemed to inspire British laundry workers to join a union.[6] The wave of organization that had swept up women workers during World War I and the early 1920s began to flatten by mid-decade. The early 1930s saw the lowest rates of membership in decades. Laundry workers' unions were also absorbed into large amalgamated organizations like the National Union of General and Municipal Workers. While the depression and the New Deal fanned activism among laundry workers in the United States, British laundry workers remained relatively apathetic.[7] Therefore, this chapter deals only with American unions. However, taken in comparative perspective, the post–Wagner Act settlement can also be read as the last chapter of American efforts to adopt the Trade Board model to American political culture. The joint boards set up by CIO unions to mediate between workers and employers were essentially private trade boards, utilizing many of the techniques familiar to reformers and union activists from state-level wage boards and the NRA.

The changes of the 1930s helped an extraordinary variety of unions flourish in the United States. This chapter contrasts two starkly different examples. Local 300 (described earlier) was characteristic of the new CIO industrial unions. Unlike some other, more closely studied CIO unions, it was distinguished not only by its diverse membership and liberal policies but also by the progressive coalition of women reformers and African American labor activists that had helped build it. In contrast, Laundry and Dry Cleaning Drivers Local 360 (AFL) was a small, Milwaukee-based union composed of white men. It derived its structure and politics from a long tradition of AFL craft unionism augmented by the legal tools the Wagner Act had provided.

Organizing Women Workers in New York

Rose Schneiderman's 1937 claim to victory was in part personal. More than any other person, she had been a witness to the prehistory of Local 300. Through the grim years of the 1920s, she and a succession of league organizers had persisted in trying to organize New York laundry workers. Her reward had most often been empty meeting halls, failed strikes, and continued betrayals by male workers and labor leaders.

"The laundries we have always with us," she wrote in 1925, while summarizing the league's organizing efforts for that year. Intensive organizing had yielded few tangible results. Her report to the league's executive board was a plea for continued patience and support. "Unionization of laundry workers can only be accomplished by a long drawn-out campaign of education," she told them. Adding a note of moral suasion, she asserted, "To stop now would be criminal, for no group of workers needs trade-unionism more than the laundry workers." Moreover, "Those who are working in the campaign are obtaining valuable experience which cannot fail to react some day to the benefit of the whole laundry industry and to the great advancement of the cause for which the League stands."[8]

Schneiderman's words were prophetic. Looking back, the 1920s could be seen as a period of preparation and learning. As in the years before World War I, organizers had been hampered by their commitment not to break ranks with the AFL. A series of organizing debacles and failed strikes revealed that both the leadership of the AFL-affiliated locals and many male members of the rank and file saw women laundry workers only in terms of their usefulness in filling out picket lines.[9] In addition, AFL organizers had particularly avoided trying to bring African American women into the union and were generally perceived as racist by nonwhite laundry workers.[10] Labor organizing during the 1920s had been a thankless task. Hostility from the courts, police brutality, and a lack of legal protections for union activities had all discouraged workers from joining unions.[11]

During those frustrating times, Schneiderman had helped build coalitions that would be essential to the membership drives of the 1930s. Contacts made as a result of a National Urban League initiative to organize laundry workers were particularly important. Under Urban League sponsorship, the Trade Union Committee for Organizing Black Workers was formed. Schneiderman attended the first meeting, along with L. Rosenthal, an organizer for Local 280 of the AFL's International Laundry Workers' Union (ILWU), and Noah Walters, an African American labor organizer who would later become secretary of the Harlem

Labor Committee in 1935 and an organizer for Local 280. Out of that meeting came a joint committee for organizing black laundry workers, which included representatives of Local 280 (a shirt ironers' union), Local 290 (a steam laundry workers' union), the Trade Union Committee, and the NYWTUL.[12]

Schneiderman resigned from the joint committee after a dismal failure of a strike in 1928. In the aftermath of the strike, she had met with Harry Morrison, the secretary treasurer of the ILWU, to air her grievances. On the top of the list was the behavior of Joseph Mackey, the district organizer, who not only displayed open animosity towards members of the New York league but also tried to coerce the male organizer hired by the league to side with the male laundry workers. Furthermore, Mackey tried to undermine the New York league's reputation within the New York labor movement. Schneiderman claimed he had "insinuated to the Secretary of the Central Trades and Labor council that the strike was illegal and that the League was responsible for its illegality." The leadership of the ILWU, she wrote to the joint committee in her letter of resignation, did not really want to have New York organized because its huge number of laundry workers "would swamp the whole International."[13] For this encounter, Mackey earned Schneiderman's lasting enmity.

The early years of the depression were indeed an "interregnum of despair" for the league and laundry workers. The ILWU counted only 6,000 members nationwide in 1929, and the New York league was in financial straits.[14] As the depression deepened, Schneiderman and the rest of the league's leadership directed the few remaining resources towards lobbying for the passage of legislation. This strategy was a pragmatic response to the crisis caused by the depression; high unemployment made organizing extremely difficult, while the league had many sympathetic supporters in New York politics.[15]

The new wave of hope brought by Roosevelt's inauguration seems to have affected workers first. In January 1934, a NYWTUL organizer reported with puzzlement a startling increase in the number of laundry workers now applying to join the union as well as a rising number of union members engaging in strike activities. She was at a loss to explain why "a group of workers that have never before responded to organization are now showing remarkable eagerness to join the union."[16] No specific evidence remains to explain what lit that early spark, but the same combination of hope, frustration, and anger fueled a nationwide transindustry strikewave that spring and summer.

Heartened by section 7A, the AFL began a new organizing campaign. Recruiting literature told workers that "the President wants you to join a union." Joseph Mackey, still district organizer for the ILWU, took up the call, again focusing his efforts on male workers. Apparently unfazed by his previous encounters with the New York league, he handed over responsibility for organizing women to Eleanor Mishnun, the league organizer.[17] Mackey's decision was probably based on the gender, not the race, of the workers and the availability of the league to do the work, but other AFL organizers made it clear that they considered African American women to be less desirable recruits than their white sisters. Two Chicago organizers told Women's Bureau investigators that they had had "a great deal of difficulty interesting white women in the union." They were frustrated that 80 percent of the applications they had received were from black women.[18]

Employers, meanwhile, had been put on notice by workers' new militancy. Laundry owners responded in a characteristic fashion, using threats and intimidation. Despite the promised protections of 7A, attending union meetings still required a great deal of bravery on the part of workers. Mishnun reported to the league that the "employers have a habit of coming to the mass meetings and standing outside the door watching who goes into the meetings." Workers who attended were later dismissed. "They also try to stop us from giving out leaflets," she further explained. In these early days of the NRA, organizers believed they could use its promised enforcement mechanisms to prevent workers from being fired. They sent NRA officials to get workers reinstated, only to find that employers used a number of ruses, such as putting workers on waiting lists, to avoid complying.[19]

By December 1933, worker enthusiasm had resulted in numerous strikes. While a number of laundry owners quickly made agreements with their workers, the owners of two large linen supply laundries in Brooklyn, the Colonial and Sunshine Laundries, refused to settle. The organizer collected pay envelopes and ascertained that most of the workers were making between six and nine dollars a week—far below the minimum set by the New York Laundry Board and the NRA. In turn, this strike prompted the walkout of workers in a number of other laundries in Manhattan and the Bronx.[20]

Infused with new enthusiasm by its membership's access to the New Deal, the New York league showed no intention of letting this opportunity go the way of so many previous laundry strikes. Believing that public sentiment and the force of the law were behind them, NYWTUL officials mobilized all their resources. Well-connected in the New York political establishment, they made sure that picketers had police protection from threatening employers.[21] After one particularly nasty incident, Mary Dreir, past president of the NYWTUL, made an appointment with the police commissioner, prompting the Chief Police Inspector of Brooklyn himself to visit the picket line at the Colonial Laundry. Soon after, the police escorts on the trucks bringing in strikebreakers were removed, and the inspector informed officers on duty that they could not interfere with picketing.[22]

The league also worked hard to court public opinion by putting out press releases emphasizing the desperate conditions of these workers and the laundry owners' willful disregard for the spirit of the New Deal and the regulations of the NRA. They called on their membership to give a public demonstration of the commitment of prominent New Yorkers to these goals. The New York membership rolls included a number of society women (sometimes known as the "mink brigade") who agreed to help. They traveled to laundries in industrial neighborhoods, stepping out of taxis and limousines to carry picket signs alongside the laundry workers. As the league had intended, these demonstrations attracted the attention of the press. Movietone News even made a newsreel of Mrs. Gifford Pinchot, wife of the governor of Pennsylvania, picketing the Quick Service Laundry.[23] These visual reports carried what was perhaps the most startling message of the strike to the public. Most of the workers with whom New York's social elite stood in solidarity were African American women. The message of interracial, cross-class cooperation resonated among African Americans. In March 1934, the NAACP journal the *Crisis* reprinted a photo of such a picket scene on its cover.[24]

Fig. 21. This photograph of striking laundry workers and members of the New York Women's Trade Union League walking a picket line appeared on the cover of the March 1934 issue of the NAACP journal the *Crisis.*

By late January, the union had still not settled with the Colonial and Sunshine Laundries, and the NYWTUL leadership again employed the social prestige of its membership to put pressure on employers. A league researcher ascertained that the laundries in question did most of their business with two New York hotels; the league mobilized its members, and the organizer later gleefully reported that "every prominent person and organization in Brooklyn and New York, not omitting the Junior League, telephoned the St. George at ten-minute intervals to inquire why they continued to patronize a laundry which defied the NRA and where workers were on strike against starvation wages."[25]

Fiorello La Guardia, the feisty mayor of New York, delivered the final blow to the laundry owners. He responded to a visit from league members by calling the employers to read them the "riot act," as Schneiderman later put it. Soon thereafter he ordered the water supply cut off to the laundries.[26] He cited $37,000 in back taxes owed to the city as legal justification for his actions, but he made it clear to the press that the real reason was the terrible labor conditions in the laundries and their failure to sign up with the NRA. In her report to the League, Eleanor Mishnun summarized the consequences of the strike: While not all the striking workers got their jobs back, those that did return received a 50 percent raise. Many other laundry owners were frightened into paying the minimum wage, and many more workers were encouraged to join the union.[27]

The complex effects of the New Deal on labor, as well as the NYWTUL's involvement with the state to ameliorate labor conditions, confirmed league members' belief that legislation and unionization were not mutually exclusive. Instead, they saw these two strategies as functioning in a complex relationship to each other. Schneiderman used a metaphor to explain this combination, arguing that it made more sense to "use both feet instead of hopping on one."[28] In the previous episode, Section 7A of the National Industrial Recovery Act and the New York State minimum wage law for women laundry workers were used to give authority and moral weight to the strikers' cause. Schneiderman was also more than willing to use legislative failures to motivate workers to unionize, suggesting that they could not always passively depend on the state.

On March 9, 1936, the New York State Court of Appeals voided the minimum wage law for women. Schneiderman and Noah Walters, her old acquaintance from the Trade Union Committee, agreed that this was a "strategic moment to begin a mass organization campaign."[29] Schneiderman counted on the anger and frustration of women who now found themselves deprived once again of legal protection. In letters to the *New York Amsterdam News,* Harlem's largest black newspaper, a number of workers confirmed Schneiderman's sense of the judgment's effect. Georgette Johnson and Marion Catherwood expressed their loss of faith in the courts and their belief that the judges on the New York State Court of Appeals were too far removed from the lives of ordinary people to understand the consequences of their decision. Mrs. Johnson addressed herself directly to the judge: "Your Honor, did you ever iron one of your shirts? How much did you think the job was worth? Next time you open a shirt fresh from the laundry, think about that please."[30]

By July, the league had helped put together an organizing committee, which included Schneiderman as secretary and Noah Walters, then secretary of the Harlem Labor Committee, as the new organizer for Local 280. In preparation, Schneiderman apparently negotiated a temporary truce between "three factional elements" in the union: most likely the male inside workers that Mackey had organized, the drivers' unions, and the female inside workers.[31]

The organizing campaign was a stunning success. Walters took primary responsibility, running the drive from the Harlem Labor Center. That location, as well as the fact that the principal organizer was black, sent an important message about the racial politics of the union. The support of the African American community and the strategies of the organizing campaign finally succeeded in allying grass-roots enthusiasm to the resources of the labor reform community. By August 1937, more than fifteen thousand workers had joined either AFL Local 280 or Local 290. In June of that year, both unions voted to abandon the AFL in favor of the CIO.

No official statement explaining the laundry workers' abandonment of the AFL survives. Undoubtedly, the move was tied to the widespread knowledge in the African American community that the AFL refused to abandon its racist policies or to discipline segregated unions. In 1941, Sabina Martinez explained that the laundry workers had decided to switch because of the "Jim-Crow policies of the high executives of the A.F. of L., which ignored the Negro people."[32]

After Noah Walters took over organization, the struggles and successes of the New York locals were celebrated in the African American press. The *Amsterdam News* put accounts of the union on the front page.[33] These stories sent the message that the union not only welcomed black members but was also run by them. Although the reality of the situation was more complex, for many African American unionists and laborers it must have been exciting to think about this world turned upside down.

Although Schneiderman worked hard to give the impression of impartiality in the battle between the AFL and CIO, some evidence suggests that she had been exploring a possible alliance between the laundry workers and the CIO-affiliated Amalgamated Clothing Workers for several years.[34] Certainly there was no love lost between her and Mackey after his repeated efforts to dump responsibility for organizing women laundry workers onto the league and then take credit after the fact. Questioned about where the New York league stood on the split and its effects on laundry workers, Schneiderman gave a cagey answer: the league was "helping both CIO and AFL unions where unions are carrying on *actual* organizing campaigns" (my emphasis).[35] Anyone familiar with Schneiderman's long battles with the AFL did not need an explanation of what that meant.[36]

Egalitarian Unionism and the CIO

Although leaving behind almost twenty years of conflict with the AFL over the status of black and female laundry workers must have been a relief, the union's strategy was risky. The now unaffiliated locals had more than fifteen thousand members and were involved in a complex set of strikes and contract negotiations. The formal, legalistic relationships imposed by the Wagner Act required the union to bring lawyers and accountants into the bargaining process. Within two months of severing the affiliation, Noah Walters and other union officials negotiated to place the former AFL locals within the jurisdiction of the Amalgamated Clothing Workers of America (ACWA), one of the founding unions of the CIO. The resulting organization was renamed Local 300, United Laundry Workers (ULW), later becoming the Laundry Workers Joint Board, which oversaw ten other New York area locals.

Schneiderman and the ACWA recorded different versions of how and why the alliance was made. In her autobiography, Schneiderman said Abraham Brickman, a young man who worked for a laundry trade association and had a social work background, came to see her about getting the laundry workers into a single organization. Schneiderman said she feared that the communist members of the local would take over in the absence of other leadership. She went to see Sidney Hillman, her longtime friend and officemate from the NRA labor board, and he agreed to approach the officers of the ACWA about including all the laundry workers in New York under the organization's jurisdiction.[37]

Meyer Bernstein, who would later head the Laundry Workers Joint Board, instead put *himself* at the center of the ACWA's various versions of the story, as published in the union's magazine, the *Advance*. Bernstein's version also emphasized the attitude of employers. He said that the newly independent CIO laundry union found itself unable to negotiate with employers in the linen and towel supply laundries. According to Bernstein, employers believed that the union leadership was potentially unreliable–unable or unwilling to enforce those contracts (this remark was perhaps a coded stab at suspected communists in the union leadership). Bernstein said he became aware of the problem while organizing the "white duck" industry (makers and suppliers of lab coats and similar garments) for the ACWA. The employers told Bernstein that they would negotiate with the laundry workers if those workers were affiliated with the ACWA.[38] Soon thereafter, the two New York locals became the ACWA United Laundry Workers Local 300 (ULW).[39] In general, the ACWA's versions ignored the prehistory of the union and any role the NYWTUL or the NUL had played, as well as the importance of Noah Walters in building the union.[40]

After the CIO affiliation, new members continued to pour in. By January 1938, the ACWA estimated that it had reached closed-shop agreements with 80 percent of New York area laundries. The trickle of one or two members that had once characterized organizing in the industry now turned into a torrent, as whole shops voted to become members of the ULW.[41]

Alliance with the ACWA provided certain material advantages to laundry workers, including organizers, office staff, and the bureaucratic support needed to coordinate such a huge organization. Local 300 had several offices, full-scale recreation and education programs, and several pages dedicated to news about laundry workers in the *Advance*. Affiliation with the ACWA also subjected Local 300 to the ACWA's version of CIO industrial unionism and put the organization under the ultimate control of ACWA leader Sidney Hillman–an arrangement that could be seen as a mixed blessing.

Hillman's vision of "industrial democracy" was of a piece with the ideas that had guided the formation of the Trade Boards and the NRA. Influenced by long interaction with progressive reformers, he believed that the ultimate goal of unionization was to stabilize relationships between employers and employees. The systems Hillman implemented to accomplish this goal were: joint boards for collective bargaining, closed shops, educational programs for workers that acculturated them to the union's goal, and a bureaucratically structured union in which experts oversaw this process. The great genius of Hillman's approach was in institutionalizing the energy and enthusiasm of a grass-roots labor movement into a form that was reproducible, controllable, and predictable.[42]

This strategy was, without doubt, extremely effective in obtaining higher wages and better conditions for the majority of workers. ACWA contracts transformed employment patterns in the laundry industry by insisting on a guarantee of eleven months of employment a year, minimum wages with overtime guarantees, and negotiation and grievance mechanisms that spared workers from face-to-face negotiations with their employers. Many employees testified to the ways that membership in Local 300 eased both the material and psychological burdens of being a laundry worker. As with the NRA and the Trade Boards, the ACWA's promise to stabilize prices and wages was appealing to employers concerned about the problems of competition.[43] ACWA-sanctioned strikes also proved to be an extremely potent tool for driving "marginal" operators out of business.

Given this commitment to stabilization and negotiation, as well as the dominance of white men in the leadership of the union, how well did the CIO live up to its ideology of egalitarian unionism? While the NYWTUL, the NUL, and the various organizations involved in organizing laundry workers before 1937 had made justice for African Americans and women an explicit part of their agenda, the ACWA alliance rendered gender and race nearly invisible. This response represented a middle ground, between the practices of some very conservative CIO unions that prioritized the needs of white male workers and the radical, communist-led unions that sometimes insisted on explicit affirmative action policies.[44] Understanding the limits of industrial democracy in the ULW is an important key for decoding the way the union rewrote its own history.

Under the jurisdiction of the ACWA, Local 300 made labor movement solidarity an absolute priority.[45] The discourse about inequality that had helped provide context and motivation for women and African

Americans to join the union before 1937 was banished. The ACWA officially opposed "discrimination on the basis of race, creed, color, political or religious belief,"[46] but it did not try to change the structural bases of inequality.

One consequence of this pragmatic stance was that, while most workers benefitted from union membership, some benefitted more than others. Examination of union contracts reveals that ACWA negotiators effectively accepted distinctions based on gender, including sexual division of labor, gender-based wage differentiation, and the notion of a family wage for men but not for women. For instance, one of the first contracts negotiated by the joint board specified wage minimums for three categories of workers. Drivers were the highest paid, at thirty dollars a week, washroom workers were to receive twenty dollars a week, and inside workers were set at a minimum of fifteen dollars a week. This scale reflected industry norms, paying the highest wages to workers who were most likely to be white men, the median wage to workers most likely to be black men, and the lowest wage to the workers who were most likely women.[47]

Fifteen dollars a week was less than most New York workers were supposed to receive under the NRA. In nobody's imaginings was it enough to live on. In the late 1930s, negotiators seemed to have concentrated their efforts on obtaining higher wages for men while hoping that a renewed New York minimum wage law for women would take care of the women's problems.[48]

In a later round of contract negotiations, the ACWA used a hypothetical set of budgets to argue that neither male nor female workers earned enough to live on. Those budgets incorporated traditional assumptions about men supporting families and women either living alone or working for pin money. The employers' representatives responded to this proposal for a higher men's wage by making a deft distinction based on the concept of a family wage accepted by both parties. The *Advance* reported to members that "wages of men in the laundry industry may be less than in other industries but laundry work does not require as much skill. . . . Also many of the 'men' are really boys with no family responsibilities."[49]

No mention was made of the extensive evidence that was available from the Women's Bureau and other organizations showing that many female laundry workers were responsible for the financial support of their families. The union was, however, willing to limit the number of hours women worked, as a concession to their family responsibilities and their presumed weaker physiques. The 1938 contract specification that women should work only thirty-six hours a week before receiving overtime, while men needed to work forty hours, might seem like a chivalrous gesture—but it also strengthened the concept of the sexual division of labor. In recognition of the importance of the family, women got time and men got money.

The reliance of union negotiators on social science data further enforced these distinctions. In 1940, the union leadership decided that vague understandings about industry norms were not an adequate bargaining tool. Therefore, the entire membership of Local 300 was surveyed. The results were tallied according to four categories: wages, hours worked, job, and sex of worker. These findings were used to create a more complex schedule of job categories used to ascertain "industry norms."[50]

The union's policies on race were less clearly defined. Unlike other CIO unions, such as the Chicago-based United Packinghouse Workers, the Laundry Workers Joint Board did not write explicit affirmative action policies into their early contracts.[51] Nor was the union used as an instrument for challenging broader social patterns of segregation and discrimination. The union did run an employment bureau, which it could have used to make a difference in racially based hiring practices. But there is no evidence as to whether racial biases of employers were discouraged or informally honored by the employment bureau. Evidence is also lacking about union members' ability to use grievance procedures to challenge racism on the shopfloor. Despite the large numbers of African Americans who belonged to the union, race generally remains invisible on the record, except in photographs in the *Advance*.

Many rank and file members received messages about their role in the union through other kinds of gender and race representations in the *Advance*. Articles were frequently illustrated with cartoons simplifying the complex matter discussed in the articles. These cartoons suggested an idealized version of roles within the union: Laundry workers were always portrayed as white. If they carried out an active role, they were male. Women were portrayed as supporters, passive bystanders, or victims to be helped. One such cartoon portrays a male figure from the ACWA protesting the minimum wage set by the state minimum wage board. A disgruntled white female laundry worker stands to the side, staring at her paycheck. The text underneath the cartoon names the actual ACWA representative to the board–she was Charlotte Adelmond.[52]

Fig. 22. This cartoon in the Amalgamated Clothing Workers' journal, the *Advance*, portrays ACWA representative Charlotte Adelmond as a white man for its readers. *Advance* (January 1938), 12.

Fig. 23. This cartoon from the *Advance* used a laundry metaphor to summarize the value of arbitration to workers. In real life, the operators of ironing machines were almost always women. *Advance* (November 1937), 12.

Another such cartoon portrays the ULW complaint mechanism as a mangle "ironing out" problems with grievances, misunderstandings, and "bosses greed." A male worker is taking the finished goods off the machine while a female worker stands to his side.[53] This kind of imagery is perhaps most meaningful in that it was explicitly intended to fire up enthusiasm among workers, to involve them in the union–yet the message it sent made distinctions between different groups within the rank and file.

Other evidence of the ACWA's essential conservatism in matters of race and gender can be found in the structure and culture of the union's management. Local 300 quickly became an example of what has often been called *bureaucratic unionism.* In structure and function, the union management replicated the gender (and most likely, the race) relationships common throughout business and government during this period. The acceptance of these norms was important because it insulated managers from experiences and perspectives of the workers. Moreover, the attitudes of management sent a message back to workers which reified those norms.

The institutionalization of social hierarchies was apparent throughout the various boards and offices that made up the bureaucratic structure of Local 300. Through articles and photographs, the *Advance* regularly acquainted its readers with individuals employed by the union. For instance, in February 1938, readers were introduced to the staff of the office that handled cases for arbitration. Each male staffer was identified according to his job. The only female member was introduced in terms of her relationship to a male staffer: she was Jack Bernstein's "alert executive secretary." Meanwhile, a month earlier the ACWA had appointed eleven organizers and twenty-one regional business managers. All were men.[54]

It is difficult to ascertain from lists of names what positions African Americans held in union management. Photographs from the *Advance* do, however, tell part of the story. Most of the faces that smile out of a 1937 photograph of the joint board are those of white males. Another photograph on the same page, entitled "They Serve the Union," shows the manager and the general organizer, Sam Berland and James Messina–both white men.[55] In 1938, the laundry workers in Westchester, most of whom were African Americans, were organized with help from the National Urban League. The *Advance* explicitly evoked racial issues in this story, describing the Westchester employers as acting like slaveholders. Yet a photograph of the executive board for the Westchester Local shows eight men and three women–all of whom are white.[56] Union member Sabina Martinez's claim that "negro women helped lay the foundation, formulate the policies and now hold executive offices" in the CIO laundry workers' union must be understood within the context of a society in which even one African American in any position of authority was cause for celebration in the African American community.[57]

Sidney Hillman himself retained control of hiring and firing employees at the highest levels of the union. His correspondence with employees and job seekers reveals a network composed mostly of the white men who had begun to make labor unions and labor politics their profession during the New Deal. It also reveals how far removed those accountants, lawyers, and businessmen were from the experiences of the rank and file. A brief example suffices to make the point.

In the late 1930s, Hillman hired a succession of managers for the Laundry Workers Joint Board. Noah Walters, who should have been an obvious candidate, nevertheless continued as an assistant manager. In 1938, Hillman hired Walter Cook as a manager. Cook was a C.P.A. and auditor who owned a small Brooklyn accounting firm. Cook had no experience with the laundry business, but he had apparently had a long connection with the business part of the labor movement. After Cook had been in office for nine months, members of the joint board asked that he be removed. In a long memo to Hillman, they cited Cook's failings: He endeavored to negotiate with laundry owners without knowing anything about the laundry business–eventually costing the union several lawsuits. He also left responsibility for negotiating the 1938 contracts in Noah Walters's hands and tried to play various members of the joint board off against one another. The coup de grâce came from Bessie Hillman, Sydney Hillman's wife and a longstanding labor activist in her own right. Hillman had invited members of the women's chorus to tea at her house, during which time Cook gave an hour-and-a-half-long speech "denouncing his fellow officers and exalting himself." Later he tried to repeat this performance at a business agents' meeting, at which point Bessie Hillman took the floor to tell him he was out of order.[58] The joint board soon had a more discreet manager.

Despite any divisive effects of such management, the union's educational and recreational programs seem to have been one of the most successful mechanisms for creating solidarity across race and gender lines. Bessie Hillman ran the educational department. An extensive effort was made to provide classes in trade union principles and skills that were open to everyone. The union also ran a summer camp program, to which both male and female union members were invited. The idea of teaching workers "trade union principles" was a longstanding progressive tradition, first carried out by organizations like the WTUL and settlement houses. Such programs had an important function in including marginalized groups of workers in the labor movement, because the claim that women or African Americans or certain ethnic groups "didn't understand union principles" was used as a gatekeeping device to prevent such workers from obtaining positions of authority in unions. Educating shop stewards and ordinary workers also took on added importance in the contractually defined world of post–Wagner Act labor unions. Members (or upper echelon officials like Cook) who acted in defiance of contracts could put the union at risk for expensive lawsuits. Education was a way to make union members' behavior more predictable.[59]

Recreational programs and entertainments also functioned to create solidarity outside the workplace. The union sponsored a band, a number of sports teams, and a chorus which was frequently celebrated in the *Advance*. Chorus members gave performances on the radio, and at least three members went on to join the cast of a Broadway musical. Photographs of the chorus illustrate its diversity: men and women, black and white, smile out of the frame.[60] Another program, the Laundry Workers Joint Board medical clinic, later touted as one of the union's proudest achievements, was not started until 1954. Access to medical care proved far more meaningful to many union members than a band or an orchestra.[61]

Whatever the identity politics of the ACWA, membership figures show that huge numbers of female and African American workers found union membership worthwhile. If the politics of the union left certain kinds of inequality untouched, these considerations were relatively minor in the context of workers' lives characterized by compromise, inequality, and day-to-day struggle. In the late 1930s and early 1940s, several African American women testified in greater detail about what they found appealing in union membership. For laundry workers Evelyn Macon and Florence Rice, union membership meant power over employers. Both made their point by contrasting their lives before and after the union and illustrating the way the union had taken power away from employers. Florence Rice described getting her first laundry job in one of the New York "slave markets"–street corners where African American women went to be hired for day work. The slave markets were condemned in the African American press as symbols of the degrading lengths to which these women had to go to get a job. "I was in the laundry before the unions were organized," explained Rice. "Twelve dollars–oh God, you had to be there at seven and I know you didn't get off at five." She added that the Laundry Worker's Union was "like the salvation."[62]

Evelyn Macon's first encounter with the CIO organizer was memorable because of the way he directly challenged the employer's right to drive workers. To the WPA writer who interviewed her, she vividly described her first encounter with the CIO organizer known as "Bruiser." He was the huge, strong man who was hired by the unsuspecting manager at the laundry where she worked. On his first day, Bruiser refused to speed up, explaining to the employer that "a man was a damn fool to rush during the first hour when he had seventeen more staring him in the face." When the boss ordered the workers to stay through lunch, eating "on the fly," he picked up his lunch and left, returning exactly an hour later. He was immediately fired, but the point was made with workers. Almost all of them attended an organizing meeting the following night.[63]

These women took a certain amount of pleasure in threatening their employers with union measures. Evelyn Macon vividly remembered her boss's fear when he discovered his workers had joined the CIO. He was "frantic," she told her interviewer. "First he tried to intimidate, then he offered to start his own union 'with the same stipulations' in our CIO contract, but we were not tricked by promises. We held our ground. After a brief strike, he called the unionized workers back to work at union hours, union wages, and better conditions."[64]

The Boys in the Vans

Halfway across the nation from Evelyn Macon, another laundry employee, with the improbable name of Kilbourn Clapsaddle, had his own reasons for being glad he belonged to a union. Clapsaddle was a laundry truck driver in Milwaukee, Wisconsin, and a member of Local 360 Laundry and Dry Cleaning Drivers (AFL-Teamsters). In October 1939, Clapsaddle was making the rounds of Milwaukee service stations, trying to solicit business for his employer, the Coverall Dry Cleaners. Among his stops was a filling station where he had previously unsuccessfully asked for business. A later arbitration report stated that he lost his temper at being turned down yet again by the owner. Clapsaddle ended up threatening that "he was going to tell fellow members of a Nazi organization" about the proprietor, implying that they would then vandalize or otherwise damage the business.[65]

The filling station owner called Clapsaddle's supervisor, who in turn advised him to call the union business manager. Had this happened before the Wagner Act, the story most likely would have ended with Clapsaddle being fired by his supervisor. Because of the contract and the union's role in providing drivers to the laundry, however, the foreman did not want to risk a lawsuit or a strike by summarily dismissing Clapsaddle on the word of someone who was not even a customer. The business manager told Clapsaddle's boss that the union would investigate whether Clapsaddle could be fired. The union disciplinary committee questioned Clapsaddle and several other witnesses and decided that Clapsaddle had indeed made "defamatory statements." Clapsaddle was instructed that he could keep his job, but he should make an apology to the filling station owner in the presence of a witness from the union. Clapsaddle failed to show up for the scheduled apology and in a subsequent hearing refused to apologize or "express his sorrow for what had taken place."

A second incident followed soon after. On a Friday evening, Clapsaddle went out to drown his sorrows with six of his fellow drivers and the route foreman. After two beers and "four or five shots," he left the Club Madrid "a little boisterous" but "not intoxicated" at around three o'clock in the morning. Unfortunately, the night's entertainment left him hung over when he reported for work the next day. "Overcome with nausea and vomiting," he left his truck in the driveway of his third customer of the morning. Someone drove Clapsaddle home and a representative of the company came and picked up the truck. Later that day, Clapsaddle was spotted celebrating his recovery by shopping with his wife in downtown Milwaukee.

At this point, Clapsaddle's employer fired him, citing violation of a company rule forbidding drivers to abandon their trucks. Clapsaddle took revenge by going around to his customers and giving them his own version of why he had been fired. He then asked for a hearing. This time, even the union could not save him from being dismissed. However, they managed to negotiate unemployment benefits and severance pay.[66] Clapsaddle soon found another job as a driver, where he continued to be the subject of conflict and arbitration. His earlier close encounter with the power of the union must have been attractive: He was later elected an officer of the union and remained a faithful member into the 1940s.[67]

If Clapsaddle was, like Evelyn Macon, a beneficiary of the Wagner Act, there was very little other resemblance between them or their unions. Local 300 was a quintessential industrial union, attracting black women who worked in the lowest-status jobs in the laundry industry. Local 360 existed for the benefit of white, male drivers. The strategies, problems, and cultures of these two locals represented opposite poles of labor organization in the industry.

Macon and Clapsaddle had almost nothing in common except union membership and their employment in the industry. The obvious differences between them were race and gender, but those qualities in turn created a third difference: class. Local 360 existed in a world of men–drivers, managers, and owners of laundries. Many of its members moved between these categories, blurring the distinction between employer and worker. In 1930s America, women like Evelyn Macon had a better chance of being hit by lightning than of owning their own laundries. Consciousness of how class, race, and gender set them apart from their employers was reinforced by daily experience and union rhetoric. For drivers like Clapsaddle, however, becoming an employer was a common and realistic goal. Aspiring drivers did not have to look far to find men just like them who had crossed the class divide. Workers at General Motors across the lake might not know the man who owned their company, but most laundry drivers did–and he was often someone very much like themselves. Reconciling this essential ambiguity with the ideology and methods of the mainstream labor movement made for some very peculiar labor politics.

This is not to say that drivers were not also exploited by their employers. As long as drivers were a burden on the payroll, employers had incentive to work them as long as possible for as little as possible. The status of drivers as semi-independent salesmen opened the door to a wide variety of abuses by employers. During hard times, drivers could be reduced to working on commission. They could also be pushed into service doing a variety of odd jobs around the laundry while other, more favored, drivers swallowed up their routes. Employers often required drivers to post bonds, which might not be returned if the laundry was bankrupted. Like other laundry workers, drivers complained of extraordinarily long hours. In a climate like Mil-

Fig. 24. Because laundry drivers interacted with customers and worked outside the direct control of their employers, they had far more power than other laundry employees. Photograph by John Vachon, 1941, Farm Security Administration Collection, Library of Congress Prints and Photographs.

waukee's, delivery work subjected men to snow storms in the winter and intense heat in the summer.

Like New York's Local 300, Local 360 prospered by offering workers and employers stability and the possibility of orderly solutions to grievances. By 1938, a substantial number of Milwaukee area laundries and cleaning and dyeing plants had signed closed-shop agreements with the union.[68] These agreements provided drivers with steady employment and a reliable wage–surely a great relief after the vagaries of the depression. In return, employers could use the union to discipline their drivers. This was a particularly valuable tool in preventing drivers from working for themselves or another laundry on the side or from taking customers with them if they went into business for themselves. While many of these problems had long been dealt with by using so-called driver's contracts, calling on the union as an enforcing agent saved employers long, drawn-out court battles.

Alois Mueller, Local 360's longtime business manager, spent a great deal of effort defining and reinforcing the boundaries between employers and drivers. In addition to taking customers with them when they left a job, drivers could also compete with laundry owners by going into business without actually taking on the expense of owning a laundry. One common method was often referred to as "bobtailing." Drivers collected laundry in unmarked trucks, negotiating a price with each customer, then delivered it to whichever wholesale laundry would offer the lowest rates.[69] Local 360 confronted a version of this problem, in the form of nearly four hundred drivers who had gone into the dry-cleaning business for themselves. Each did all his own work and then delivered it himself. Much to the disgust of the Teamsters (who insisted these budding entrepreneurs be forced to join Local 360), the AFL International Laundry Workers' Union tried to organize these drivers into an employers' union.[70]

Local 360 also carried out some of the functions of a traditional trade union by regulating competition between drivers. It divided drivers into categories and mediated territorial disputes. Some union drivers manipulated the old and new dispensations to secure new business. In at least one case, union drivers tried to convince their competitor's customers to switch laundries by hinting that workers in the competing laundry were about to go on strike. This tactic definitely fell under the Wagner Act rubric of "unfair competition," and the offending drivers were fined by the union for their cleverness.[71]

Union contracts removed many elements of the employer's ability to hire and fire workers at will. As labor leaders rightly argued, this provision protected workers against capricious employers. Contracts could also have more peculiar effects, deeding power over to union officials or allowing that power to function in arbitrary ways. Local 360's standard contract obliged employers to hire only union members in good standing and to fire workers who did not pay their union dues—in effect, making employers responsible for dogging their employees about late dues. Because hiring a new driver who had a poor relationship with the union could endanger the contract, many laundry own ers relied on Alois Mueller to recommend drivers when jobs came open. Since Mueller was privy to all complaints, he should have been able to warn employers against unreliable drivers. In practice, however, being an active union member and cultivating a good relationship with Mueller might simply be a driver's way of compensating for a problematic employment history. The aforementioned Clapsaddle never had any trouble getting a job, despite his various transgressions.

Because specific laundry drivers became, in the minds of consumers, synonymous with certain laundries, either a driver's bad behavior or the loss of a favored driver could be extremely damaging to a laundry's business. The union business manager's power to remove offending workers from the union, and therefore from their jobs, was terrifying to some employers. One frightened owner of a small family laundry pleaded with Mueller not to discipline their driver for being behind on dues. "The way business is today we absolutely cannot afford to lose a dollars worth of business, and our past experience has taught us with Oscar off of the truck it means a decline of from $20 to $30 a week. We assure you that you will have no further trouble with either his dues or his conduct in the future."[72]

Savvy or fearful employers worked hard to cultivate a good relationship with Alois Mueller. Oliver Adelman, owner of Adelman Laundry, was in the habit of sending Mueller clippings from industry trade journals, since, as he put it in an accompanying note, Mueller had "the interest of the laundry industry at heart." Adelman also Anglicized Mueller's Germanic name to Al Miller, perhaps in a subconscious effort to make him into a more palatable colleague.[73] Other laundry owners tried appealing to a higher authority to subvert Mueller's power. Howard Fink, the owner of the Sweet Clean Damp Wash Laundry Company, wrote to Julius P. Heil, the governor of Wisconsin, complaining that Mueller (also "Miller" in his complaint) had tried to dictate what prices his laundry would charge for certain services should he open a plant in Milwaukee. Mueller responded to the governor's inquiry by stating that Fink's accusations were untrue and that, in any event, he did not believe that the matter was "one which requires the attention of any outside individual, official, or governmental agency." Mueller was excused from any wrongdoing and eventually cultivated a congenial relationship with Fink over lunches and dinners. Like Adelman, Fink continued to address Mueller as Al Miller.[74]

Mueller and his counionists nevertheless continued to use both traditional methods such as the strike and strong-arm techniques to expand the power of Local 360 in Milwaukee. The union was often willing to ignore drivers who crossed back and forth between proprietorship and being employees, as long as they continued to pay their union dues and were willing to follow "union principles." In practice, *union principles* encompassed not only the specific rules and conditions laid out in the union's charter, added resolutions, and contracts, but also conventions of behavior. Members used terms like *disobedient* and *disloyal* to describe behavior that overstepped these often unwritten boundaries.

Occasionally Mueller chose to make an example out of a driver who was thought to be publicly challenging this unspoken agreement. William Stadler left his job as a driver to work for his wife, who owned a cleaning and reweaving business. Soon thereafter, the union passed a resolution forcing employers to ask for minimum rates on wholesale accounts (ensuring a minimum commission for drivers). Stadler protested because the resolution affected his wife's business, and he was promptly suspended from the union. Mueller apparently still saw Stadler as a threat to the union, even after having suspended him. According to the Stadlers' lawyer, Mueller had arranged to have Mrs. Stadler (who was at the time in a "delicate condition") and her husband followed and observed. He also sent damaging letters to Mrs. Stadler's customers. Mueller's attempts to force the Stadlers to cooperate eventually led Mrs. Stadler to hire a lawyer, who argued that the business did not legally fall within the terms of the resolution. The lawyer further argued that the Stadlers were "carrying on an honest, decent business in a lawful manner and in a manner consistent with union principles. They are trying to make a decent living for themselves, their child and the people to whom they can give employment."[75]

Sometimes Mueller's efforts to use extralegal means to discipline disobedient union members backfired. Mueller somehow managed to get Walter Rothenhoefer, a driver who was bobtailing on the side, arrested for drunken driving. The judge apparently let Rothenhoefer off the hook despite a blood alcohol level of .32 percent. Mueller was so angry that he consulted with the union's lawyer about sending a letter to the Milwaukee Journal complaining about the judge's actions. The lawyer advised against it: "I do not think it advisable to make an issue of Judge Neelen's attitude in this case. You must remember that we, too, sometimes ask the judge to blink at the letter of the law." The union ended up paying Rothenhoefer a settlement for unlawful arrest.[76]

Meanwhile, many drivers simply paid their dues and otherwise ignored the union. For others, fellow union members provided a form of community. Some derived a sense of empowerment from participating in union affairs. Others simply found the union hall a congenial place to spend some of their free time. Local 360 lacked the elaborate educational and recreational programs of the ACWA. Mueller seems to have assumed that trade union principles would come naturally to white male union members, that trade union culture was contiguous with the culture of working-class masculinity. In a sense, he was right—Milwaukee had long been steeped in the politics of the labor movement. Mueller also simply chose to dismiss any ignorance of members as disinterest or defiance.

Mueller was no fool, however. He was acutely aware of the importance of entertainment in keeping members interested in the more serious business of the union. While Local 300 offered its membership free healthcare, an annual dance with the Count Basie Orchestra, and tea with Bessie Hillman, Mueller tailored the Wisconsin union's benefits to its very different constituency. For instance, no union meeting or other event took place without a liberal offering of beer and pretzels. In 1938, Mueller sent out a form letter inviting employers and members to a meeting discussing jurisdictional disputes between the AFL and CIO. He seemed little worried about the results, telling the invitees: "Regardless of the outcome, whether it be A.F.L. or C.I.O, we are going to throw a little party, and have plenty of beer and lunch for those who attend." By 1939, Mueller had bought so much beer from the City Beer Exchange that he felt confident in dickering over a price increase.[77] For a stag party later that year, he paid $150 out of union funds to a theatrical agency for a variety revue. Drinking seems to have been a more successful draw than the union's bowling league, which quickly collapsed.[78] Union meetings accommodated not only beer and pretzels but also other diversions of masculine culture. A vote was taken during a February 1940 meeting to suspend the regular order of business in order to listen to a boxing match broadcast on the radio.[79]

Eventually, the exclusively masculine atmosphere of the union's social events must have worn thin with its married members. In 1941, the entertainment committee sent out an invitation to members' wives, apologizing for not including them in union entertainments and inviting them to a showing of two movies: Fred Waring and his orchestra in *Pleasure Time* and a short called *Tobacco Land U.S.A.*, with refreshments and dancing afterwards.[80]

Wartime labor shortages soon rocked the all-male world of Local 360. Alois Mueller joined the Navy. The Teamsters decided to issue a special membership form for taking in female members during what they termed "this emergency."[81] Union officers did their best to help women like Dorothy Fischer and Elizabeth Mauer, two of the female drivers who joined the union during the war. Mueller's replacement, George Scanlon, even fought for Mauer to keep her job after the war on the basis of her seniority rights.[82]

✖

By the late 1930s, unionization and regulation had become a pervasive influence in the affairs of many laundry owners and laundry workers.

In Britain, the occasional entrepreneur, friendly with a local magistrate, might avoid paying the minimum wage to learners or extract a little illegal overtime, but this was an exception. Because of the federal system in the United States and the more important role of unions there, region played a larger role in defining who would be affected.

The dominance of linen service workers in unions like Local 300 was also a harbinger of the future. In the same years that labor reconstituted its role in the industry, the housewife-consumer, so central to laundry owners' efforts to expand the industry, began abandoning laundry service, leaving laundry owners more and more dependent on commercial customers.

11

Facing the Domestic Threat

WHAT HAPPENED TO FINALLY SET THE LAUNDRY INDUSTRY INTO decline? The simple answer often provided is that increasing numbers of consumers chose home washing machines rather than laundries.[1] By sometime in the 1950s, the deed was done. Whereas in 1931, family services had made up nearly 80 percent of receipts, a much diminished postwar industry hung on mostly in the form of diaper and linen services.[2] Dwellers in tiny Manhattan apartments and residents of swank hotels might habitually send clothes to a laundry, but most people without washers relied on the dry cleaner or the laundromat (or *laundrette,* as the English call them). Judging from numbers of washing machines per household, several historians have also argued that this process was slower in Great Britain than in the United States, leaving the British laundry industry relatively undisturbed until after the war.[3]

While the end seems certain in retrospect, the process through which it came about illustrates the continued complexity of the relationship between technology and culture in shaping the industry's fortunes. Washing machine manufacturers and laundry owners fought for the householder's patronage over the course of nearly three decades, though with a decided interruption during the war. Those battles (the military metaphor was evoked by laundry owners themselves) ranged across a terrain rocked by other transformations.

Among these changes, shifts in consumer culture directly affected the outcome of the struggle. For instance, washing machine manufacturers benefitted far more than laundry owners from the spreading availability of consumer credit. Widespread electrification sped the adoption of washers in the United States. Piecemeal electrification and trade barriers against cheap American machines slowed British purchases. Without doubt, Americans bought far more washing machines than the British during the interwar years. Only about 150,000 electric washers were in use in Britain by 1939. This low number is partly explained by the alternative use of the much cheaper washboilers in Britain—a device unknown in the United States. Nevertheless, the total number of washers and washboilers in use at the end of the 1930s in Britain has been estimated at only 300,000—meaning that less than five percent of British households had either appliance.[4] In contrast, American washing machine manufacturers sold 1.5 million machines in 1929 and estimated six million were in use in American households that year.[5]

More washing machines in America did not mean the American laundry industry lost ground more quickly than its British counterpart. During the 1920s, the British laundry industry grew more slowly than its American counterpart, despite less competition from washing machines.[6] Rising wage rates, caused in part by the Trade Board Act, and economic depression had begun to affect British laundry profits during the early 1920s. Given the rarity of electric washers in Britain during this period, it is likely that consumers chose not between the washing machine and the laundry but between the laundry, the washboiler, the laundress, and doing the work themselves using a copper and a dolly.

During the prosperity of the 1920s, Americans bought a large number of washing machines, but they also dramatically increased patronage of laundries. Just because someone bought a washing machine did not necessarily mean a laundry lost a customer. Some machines were bought by people who had not patronized laundries before. In addition, many washing machine owners still sent hard-to-wash items to the laundry. The Great Depression initially sent Americans back to the washerwoman and the tub, but by mid-decade, washing machine manufacturers had figured out ways to make mass production and consumers' economic anxieties work for them. The 1939 census revealed the laundry industry was also still growing, but at a much slower rate than it had in the 1920s.[7]

Initially, the real casualties of successful washing machine sales were washerwomen. Or, alternatively, housewives bought washing machines because they could no longer find other women willing to do their laundry cheaply. In slow decline since 1910, when census takers estimated their numbers at 520,000, the ranks of women identified in the census as "laundresses, not in laundry" thinned dramatically during the 1930s. A decade later, the vast majority of the 189,592 remaining laundresses could be found mostly in Southern states.[8] Washerwomen themselves may also have been among the purchasers of washing machines. One washing machine dealer in the 1930s suggested to fellow dealers that they search the want ads for washerwomen advertising to take in washing, then call up these women and convince them to buy a machine to improve their businesses.[9]

Choosing a Washing Machine

A complex combination of economic, technological, and cultural factors motivated women's abandonment of commercial laundry services. New ideologies of domesticity, the feminization of consumption, and manipulation through marketing certainly played a role, but women were not passive dupes throughout this process. Women who had the resources to choose between alternatives may have been swayed by advertising, but they also drew on their own experiences and weighed their own needs and desires when choosing to use a washerwoman, a washing machine, a laundry, or some combination of the three. As Chapter 6 has suggested, dealing with a laundry was often enough not the carefree exercise LNA advertisements made it out to be. It saved time but not always aggravation. Since many consumers had already abandoned expensive finishing services in favor of rough dry work, they were already committed to spending some time doing laundry.

In May 1932, the British trade journal *Laundry Record* published an editorial entitled "Grave Threat." The author advised other laundry owners that "vilification of the domestic washer will do no good" in protecting the industry against competition. With more candor than literature on the subject usually allowed, he stated, "The claims of the salesmen are not exaggerated; interviews with housewives confirm that the domestic washer is efficient, easy to run, cheap in operation, and very durable." Indeed, by the 1920s washing machines had become, if not the self-minding devices of today, easier to use than their predecessors and much less labor intensive than the washboard and the dolly. Adding a source of mechanical power to supplement the strong arms of the laundress made a critical difference. Ultimately the electric motor proved the most successful solution to this problem, although dealers did a steady business in machines with gas engines through the 1930s.[10]

The technology of home laundry work consisted of a system of machines or processes dependent on a larger technological infrastructure. Laundries had subsisted and often prospered by being self-contained. In the context of nineteenth-century cities, steam engines, wells, and water purification systems gave laundries an advantage over householders who might or might not have access to water and could rely only on the force of strong arms, servants, and children to drive household technologies. The spread of water and power systems–and the democratization of access to these systems–changed all this.

Electricity was one important component of that process. Electrification spread much quicker in the United States than in Great Britain. In the United States, a common electrical frequency was introduced in most locations in the few years after 1900, thanks mostly to the efforts of the Westinghouse company.[11] This urban electrification took place most rapidly during the 1920s: one American home in seven had electricity in 1910, but by 1930 the ratio was seven homes out of ten.[12] There was actually a slight drop in the number of electrical meters during the early years of the depression, as some newly electrified customers went back to kerosene lanterns and woodstoves to save money.[13] The Rural Electrification Administration of 1935 also brought in a host of new customers. Unlike their urban cousins, rural housewives often saw a washing machine as a first-priority purchase after electrification.[14]

Meanwhile, British electrification lagged behind that of the United States. A national standard was not introduced until the British National Grid was set up in 1926, gradually replacing a patchwork of local systems of varying frequencies and voltages. Thereafter, Britain electrified at a remarkable rate. In 1920, only one house in seventeen was wired for electricity in Britain. By 1930 that ratio had risen to one house in three, and by 1939 it was two houses out of three.[15]

Electrical standardization made it far easier for washing machine manufacturers to produce goods for a national market. Beforehand, in order to use appliances consumers had to buy separate motors appropriate to the voltage of their local electric systems. The first "all of a piece" electric machines began appearing on the American market between 1910 and 1920. Consumer studies in this period found, however, that these machines were quite rare and owned almost exclusively by the upper classes.[16]

Lack of standardization forced British appliance manufacturers to gear their efforts towards highly localized markets, a trend that continued well into the 1930s. With the introduction of a national grid, British manufacturers found it difficult to compete with the cheap mass-production techniques of their American rivals. Although manufacturing of electrical appliances for British use was also part of the "new industries" that gave a postwar boost to the economy of the Midlands, their market share was small compared with the transatlantic firms of Hotpoint and Thor. Britain's abandonment of the gold standard, in 1931, followed by the 1933 Import Duties Act, dammed this tide of foreign imports and pushed British consumers to buy domestically produced machines during the 1930s.[17]

In the home, as in laundries, washing machines were part of a technological process that also required other devices. Electrical appliance manufacturers had already made their first inroads on the domestic laundry market with electric irons, not washing machines. From the early twentieth century, these irons had been one of General Electric's biggest sellers. Their obvious superiority over the sadiron led thousands of consumers to trade in their old irons for the new electric appliance.[18]

To compete with commercial laundries, some appliance manufacturers believed that they had to offer machines that could do "the whole job." Beginning in the late 1920s, manufacturers like General

Electric began offering a variety of other substantial devices modeled after commercial laundry equipment. Chief among these was what was generally called an *ironer*. Modeled after the mangle, it was a single padded roll that could be used to iron flatwork and (if one believed the manufacturers) nearly everything else as well. Unlike the hand iron, an ironer was a substantial investment, costing nearly as much as a washing machine.

Ironers never became big sellers. In 1930, *Electrical Merchandising* estimated that there were only 580,000 ironers in use in the United States, compared with nearly 7 million washing machines. Not only was the high cost of such devices off-putting, but consumers found their clumsiness and lack of portability unappealing. Also important in explaining the greater number of washing machines was the introduction of the hot water heater, obviating the need to solve the technically difficult problem of heating the water inside the washing machine.[19]

While electric washers were an improvement over their predecessors, they still had many drawbacks in terms of the labor and danger involved in using them. Ad copy that implied one could push a button and leave for the afternoon was misleading. Early machines did little more than move water through the clothes. Hoses had to be attached to fill the machines and drains opened to draw off the dirty water, a process that had to be repeated for each rinse. That awful affliction, "tattle-tale grey," was largely caused by failing to use an adequate number of rinses to get all the soap out of the fabric. Advertisers were not far off the mark when they suggested that it was a labor of love to get clothes white and bright. Those extra rinses represented a substantial investment in time and labor.

Mechanical failures were frequent in the earlier machines. Motors sometimes shorted with exposure to water, and they had to be regularly lubricated so as not to freeze up. Frayed cords and faulty engines also contributed to the risk of electrocution.[20] Parts of the process that had been mechanized in commercial laundries defied easy adaption for the home. Commercial centrifugal extractors, for instance, ran at high speeds that strained moving parts and put the user at risk. To prevent these extractors from moving, laundries bolted them to the floor—an unacceptable solution for many homeowners. Consequently, rather

Fig. 25. Doing the family laundry with early washing machines was time and labor intensive–far from the glamorous exercise portrayed by some manufacturers and women's magazines. Photograph by Ralph Andursky, 1943, Rochester, Farm Security Administration Collection, Library of Congress Prints and Photographs.

than an extractor, most washers had a wringer mounted on the side of the machine to remove excess water. Electrically driven wringers were sometimes the source of gruesome accidents, catching the hair or clothes of the unvigilant–a danger laundry owners used to great effect in their anti–washing machine ads.[21]

She Bought One! . . . .

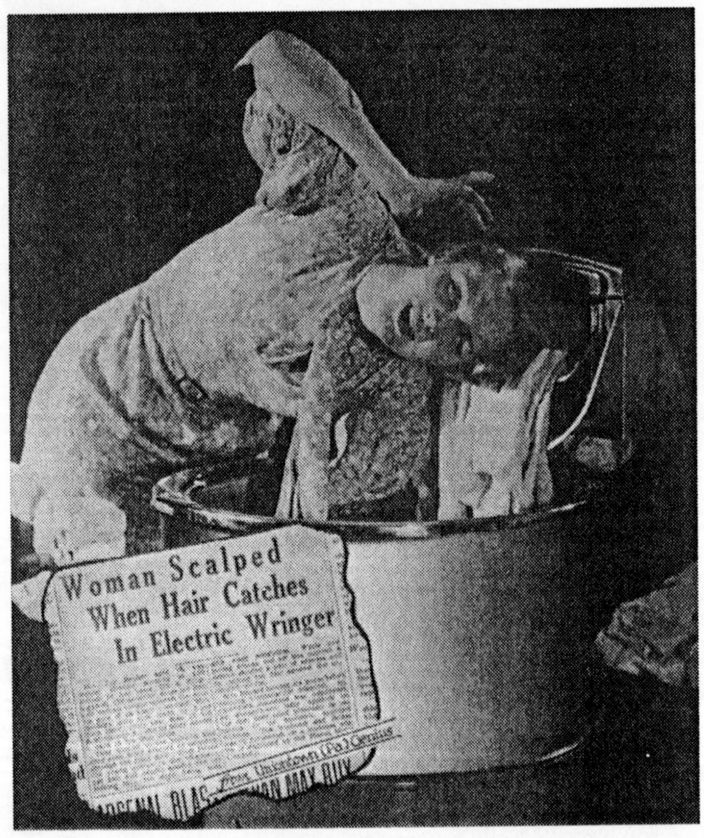

Fig. 26. Laundry trade associations employed scare tactics in an effort to prevent their customers from buying washing machines. *Laundry Age* 15 (April 1935), 9.

Some manufacturers did offer an extractor as a separate unit attached to the washer. These machines had the same problem as their commercial counterparts: they were prone to becoming unbalanced, causing the machine to dance across the laundryroom floor and wear out its bearings. Manufacturers solved this problem partly by making extractors with very small load capacities, providing careful instructions about how to pack them, and attaching extractors to heavier washing machines. Extractors also drove up the price of washing machines, however, so most consumers continued to make do with a wringer.

As with all new technologies, people had to be instructed in the proper use of home laundering equipment. Instructions on operating machines often included a simple reminder of the necessity of plugging the machine into the electrical wall socket. Home and store demonstrations, instruction manuals, and magazine articles aimed at women consumers all taught technological knowledge while also encouraging consumption.[22]

Even with electrification and improved technology, doing the wash at home took time and money, as the laundry industry liked to point out in its advertisements. Home economists struggled with the question of whether washing at home was more economical. After factoring in an hourly wage for the housewife, results were inconclusive.[23] However, most housewives made consumption decisions within the context of a family economy. Many middle- and working-class women worked with limited allowances doled out by their wage-earning spouses. Doing the work oneself freed up money for other purposes. Even laundry owners admitted that doing the wash at home involved a lower cash outlay.[24]

During the 1920s, dealers learned the fine art of making appliances like washing machines seem more affordable, and therefore competitive with the cost of sending clothes out to the laundry, by offering consumers credit on a time-payment plan (or "hire purchase," as it was called in Britain).[25] With the onset of the depression, dealers resorted to very low down payments—as little as 75 cents in one instance—and credit plans that spread payment over a period of years to attract customers.[26] British practices mirrored their American counterparts. The editor of *The Power Laundry* warned his readers that department stores and light and power companies were offering electric appliances for hire-purchase with payment periods as long as five years.[27]

Particularly in the United States, rapidly decreasing prices coupled with such payment plans made washing machines affordable to a wide range of consumers. In 1925, a moderately priced machine cost between $125 and $150 in the United States. In Britain, machines with a separate electric motor cost between 30 and 50 pounds during the 1920s (slightly more than the American machines).[28] In the early 1930s, Sears and Roebuck initiated a price-cutting war that targeted its competitor Montgomery Ward but affected all dealers. By 1934, the bottom-of-the-line Sears washer cost $29.85. Other, more sophisticated machines could be had for between 50 and 60 dollars.[29] In the absence of such a price war, prices fell only slightly in Britain over the same period.[30]

While Sears could afford to swallow a temporary loss to attract new business, other American retailers panicked at the narrow (or nonexistent) profit margins that such prices entailed. By the same token, consumers recognized a good thing and business boomed. In 1936, Americans bought 1,533,300 electric washers and 199,700 gas power machines, and dealers began to see large numbers of return customers as well.[31]

Perhaps paradoxically, the depression gave washing machine manufacturers an advantage over American laundry owners. Common sense suggests that people would be unwilling to buy high-ticket consumer durables during a depression, but several factors counteracted this tendency. While price wars among laundries ravaged the industry, sending laundry owners looking for government stabilization, price wars among washing machine manufacturers favored a few large firms with the assets to survive temporary losses and the capacity to produce cheaply by taking advantage of volume. Through this process, companies like Maytag attracted large numbers of new customers and gained a relative monopoly in the market.

At least one dealer also suggested that consumer durables offered a peculiar but safe form of investment in an era of unstable financial institutions. In 1934, he argued that "the American housewife for the next few years will bank the family savings in comforts for her home; she's learned a lesson about placing a rainy-day reserve in banks, in stocks, and real estate. . . . Home appliances are the one investment which pay for themselves, cannot be frozen or destroyed, and which can be used and enjoyed no matter what happens."[32] Moreover, unlike a stack of old laundry lists, in the most desperate of circumstances machines could also be sold to another dealer, recouping some of the original cost and feeding the secondhand market.

One did not have to own a washer to use one. Technological innovation allowed washing machine manufacturers to carve out a niche among some of the commercial laundry's most consistent customers: apartment dwellers. Coin-operated machines were available from the early 1930s. Entrepreneurs entering the coin-op business dealt with the predictable problems attached to a new technology. The first coin boxes were easily tampered with, and the washing machines themselves were small and relatively lightweight compared with later models. A reasonably strong individual could move one without difficulty (as the salesmen who moved them on and off of trucks for home demonstrations knew). Savvy operators chained the machines to walls or pipes to prevent theft–a danger that increased on the national moving days, May 1 and October 1.[33]

As always, notions of cleanliness also played a role in consumers' choices. The 1920s and 1930s are marked by two significant changes in this respect: Energetic advertising, particularly on the part of soap manufacturers, accelerated the long-term trend of increasing standards of cleanliness. Advertising copy suggested that cleanliness was not necessarily self-evident–sure, "tattle-tale grey" was a problem, but so were invisible dirt and germs. As the *American Home* magazine explained, it is not the danger of germs which "prejudices the fastidious homemaker and her family against mixing personal clothing, table and bed linen with that of utter strangers"–but rather the uncomfortable feeling of someone else's presence. As one woman put it: "I can stand my own dirt but I can't stand the dirt of someone I never saw."[34] This notion of the commercial laundry as the repository of other peoples' dirt and the dirt (in both the literal and symbolic sense) of workers, including an increasing number of African Americans, coupled with another cultural change–the increasing demand for privacy and a clearer delineation of public and private–gave consumers vague twinges of distaste at the idea of strangers handling and inspecting their private things.[35]

In 1936, the British journal *Electrical Age for Women* carried an advertisement for the British-made H. M. V. washer and ironer that neatly summed up the industry's arguments for choosing a washing machine. A cheerful housewife looks up from her work to announce, "I've finished paying laundry bills, and 'for good.'" She claimed to spend only an hour washing, with better results and less damage than could be obtained from a commercial laundry. "Besides," she adds, "I never really fancied the idea of my laundry going through the wash with everybody else's!" The H. M. V. washer was so easy to use, she said, that she "could do the weekly wash in a dance frock!"[36]

The Battle Is Joined

Just as consumers were not passive in making choices, manufacturers and laundry owners played active roles in trying to shape those choices. During this period, they engaged a bitter battle for the hearts and minds of both British and American consumers. Inch by inch, year by year, the laundry owners lost ground, done in by their own strategies as well as the broader terrain that constituted their battleground.

It was a battle in which laundry owners sometimes proved the agents of their own undoing. They often failed to confront the nature of the new technology's appeal, choosing instead to view the context for laundries as having remained unchanged from the early years of the century. In addition, although the LNA kept well abreast of new developments in the marketing and management of laundries, many smaller laundry owners continued the haphazard approach that had always characterized the trade.

Particularly threatening was the centralized nature of the washing machine manufacturing industry and its ability to harness the power of that transforming force of twentieth-century life, advertising. By the 1920s, the majority of washing machine manufacturers were based in the state of Iowa, and giants such as Maytag were busy creating national and international markets for their products.[37]

The common-sense notion laundry owners depended on to sell their services–that home laundry work, even with a machine, was difficult, time-consuming, and hardly more economical or sanitary than that done by the laundry–was under constant bombardment by the machine manufacturers. While laundry owners were no strangers to advertising, the localized nature of the industry (even under the auspices of the LNA) precluded the kind of nationwide, saturation advertising that General Electric or Maytag could afford. Both the LNA and the Federation of Launderers relied on membership dues and contributions to pay for campaigns. The weaker federation found it particularly difficult to convince its members to make contributions towards a national publicity campaign against washing machines.[38]

The LNA and washing machine manufacturers waged an advertising battle so bitter that two truces had to be called in 1927 and 1933 to cut down on the amount of negative advertising.[39] Women's magazines provided the main forum for advertising competition. Because washing machine makers (along with soap manufacturers) provided such a large percentage of advertising revenues, editorial policy was often slanted in favor of home washing.[40] Some examples seem almost like parodies of the genre. In 1937, *Good Housekeeping* featured an article titled "Glamour in the Laundry! Joy in Homemaking!" One of the author's informants declared her joy when pulling clean clothing out of the washer or ironing: "I feel as an artist must feel as his picture unfolds on the canvas. . . . There's glamour in it for me."[41]

Women's magazines also worked to influence consumer choice through gimmicks such as appliance testing and the Good Housekeeping seal. In response, the LNA created a central research and training facility similar to the Good Housekeeping Institute and a system of inspection. Successful laundries were given the right to use a seal of approval bearing "the smiling face of a white-capped girl" and the legend "Approved by the American Institute of Laundering."[42]

Machinery manufacturers built on the domestic ideology of the 1920s that emphasized the home as a display place for consumption. They encouraged the idea that the home washer was not only a useful device but also a status symbol. Appliance merchandisers latched onto and made extensive use of the phrase *proud owner of a* ––, while *Good Housekeeping* asked, "Are You Proud of Your Laundry?" The text emphasized the importance of owning a full range of "up-to-date equipment."[43]

The British appeal to consumer pride was couched in subtler terms. The text for one British ad emphasized the pride a housewife might take in the good sense displayed by the purchase of a washer: "She sneaked away to wash! She did not want it seen, nor did she want it known that *she* did the family clothes," but, the ad continues, "TODAY the sensible woman takes pride in doing her own washing. She calls in the neighbors to see how little work she does and how much the UNIVERSAL electric washer does for her."[44] The text also emphasizes another important social effect of the technology: in the acutely class-conscious atmosphere of British society, the use of a machine lessened the stigma of doing the work oneself because it implied some kind of expertise in both operation and consumption.

British arguments for adopting electrical appliances also contained an element completely absent in the American context–a fear of British technological backwardness. In 1926, Wilfred Aster told members of the Electrical Association for Women about her recent visit to Canada and the United States, where she had glimpsed the possibilities of electrical liberation. No Canadian or American woman would "dream of using her own muscular energy in washing, ironing, or sweeping," she told her audience. The Canadian woman loads her washing machine, "presses a button and puts the lid on, and goes away to attend one of these lectures or one of those very attractive concerts at 11:00 and comes back to find her washing done."[45]

Manufacturers also emphasized what they termed *emotional selling*. These strategies depended on emphasizing family values and the clear definition and delineation of gender roles. The August 1934 cover of *Electrical Merchandising* promoted these methods: the cover featured a bedraggled-looking woman straining over a washboard in a bucket of suds. The copy read: "Emotional Selling–to free women from inhuman drudgery–to safeguard the health of children . . . these are the results of electrical appliance merchandising."[46] Some strategies bordered on the ridiculous. The Electrical Merchandising League of Cleveland ran a series of columns in local newspapers modeled on the popular "advice to the lovelorn." "Peter Fairfax's Letters from Lovers" told newspaper followers how to (literally and figuratively) "iron out domestic wrinkles" through the use of domestic appliances. "Romeo Capulet" wrote

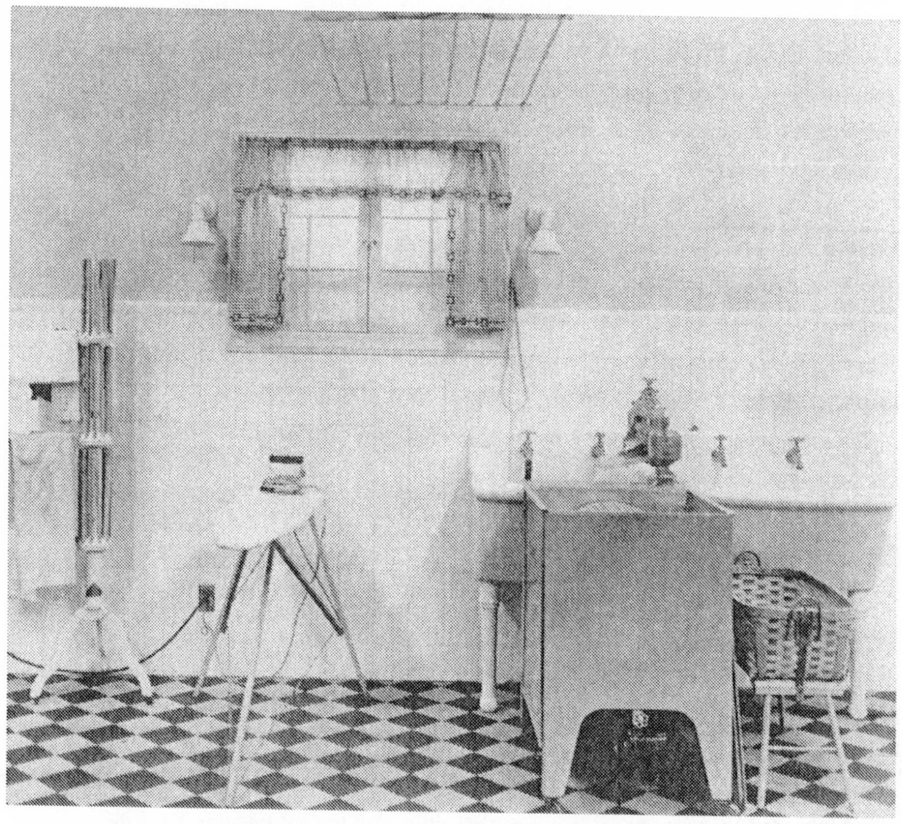

Fig. 27. This model laundry room, equipped with the latest technology, represented the kind of ideal with which merchants and manufacturers tried to entice new customers. Library of Congress Prints and Photographs, 1924.

to complain that his marriage seemed destined for disaster because of his bride's mysterious irritability on Mondays. Fairfax of course advised the purchase of an electric washing machine. Similar relationship problems could be solved through the purchase of irons, toasters, or waffle makers.[47]

Such approaches were figured strongly in the war over customers. Advertisements for laundries suggested that home work put an undue burden on women's time and energy. Washing machines wore down women's strength and distracted them from the more important matter of attending to their children and husbands. They were also unsafe, reaching out to kill or maim the unvigilant user or innocent child.

Machine dealers and laundry owners both understood marketing in gender-specific terms. Tactics were often geared to the supposed "nature" of women.[48] With remarkable frequency, the idea that women wanted washing machines because they saved labor seems to have gotten lost in stereotypes of women as acquisitive, emotional, frivolous, and houseproud. One piece of advice to retailers about a springtime ad campaign to sell appliances suggested that "this is the time when the young homemaker wants to 'love, honor, and display.'"[49]

Because men (specifically as husbands) most often made the final decision on the purchase of big-ticket items, sometimes the relationship between retailer and consumer took on the form of a triangle. "Women initiate, men decide," began the text of one article on selling washers. "She must be turned into a salesman on her husband. If we give her the arguments, she will give him the works."[50] Some retailers went straight for the husband. One Toronto store set up a window display featuring a tombstone emblazoned with the text: "Here lies the body of Hiram Green; her husband wouldn't buy her a washing machine." Another dealer dubbed his shop the "Wife-Saving Station."[51] A machine dealer in Lorain, Ohio, employed a more sophisticated, cross-cultural approach. He created a window display targeting male workers of this largely immigrant community. Its text suggested that real American men spared their wives the tortures of washday by buying appliances.[52]

Ironically, the isolation created by the restricted environment of housework offered another opportunity for appliance sales. Dealers lavished attention on housebound women, with phone calls, rides down to the stores, and social corners in stores where they could have a cup of tea and consider the possibilities of future purchases. Some laundry owners retaliated in kind. The 3-F Laundry of Madison, Wisconsin, tucked an edition of the *Bundle News* into each batch of clean laundry. It featured stories and articles unrelated to the laundry, recipes, advice on childcare, and a liberal sprinkling of advertisements emphasizing the value of laundry services.[53]

A wide variety of organizations and institutions arrayed themselves behind the effort to sell Americans on home washers. While laundries had no natural allies among outside industries and organizations, washing machine manufacturers could count on the backing of electrical companies and, in some cases, even government agencies. These alliances spawned a secondary level of organizations aimed at finding ways to reach the consumer. In Britain, the "Women's Electrical Association" was formed because of "the need to interest women in regard to electrical development." In practice, it became a means of convincing women to buy and use electrical appliances.[54] Various merchandising associations in the United States, as well agencies of the United States government itself, all lobbied consumers to increase their use of electrical appliances.

⁙

The choice of a home washer over laundry services should not be read as a return to a preindustrial past. While laundries had a great deal of difficulty detaching themselves from their nineteenth-century technological and cultural roots, washing machines fit neatly into the developing structures of twentieth-century industry and business. This in itself was no formula for the laundry industry's failure. However, the strategies and shortcomings of the industry became increasingly problematic as consumers found they had alternatives. In the interwar period, no amount of careful organization and better machinery could completely disguise the laundry industry's awkward adoption of factory methods while still keeping services affordable. Most laundries were also local businesses. Despite educating, proselytizing, and occasionally coercing, laundry trade associations could not guarantee the same degree of standardization that could be built into a brand name machine.

In contrast, the electric washing machine was a quintessential twentieth-century technology. A Maytag was a Maytag whether one bought it in Des Moines or Los Angeles or Liverpool. The most successful washing machine manufacturers were big businesses that used a variety of techniques–mass production, national and international marketing, and consumer credit–to sell their products.

It is likely that the high price of washing machines, the unevenness of electrification, and the greater availability of washerwomen combined to slow the adoption of washing machines in Britain. In the middle years of the twentieth century, new characteristics of industrialization found their mark in the United States first. A British laundry engineer returning from a visit to the United States in 1935 gave his fellow laundrymen a different reading of the situation. "Britain Now Leads" trumpeted the headline of a trade journal article. He ascribed the American crisis to overconfidence; American laundry owners had expanded too quickly, with too many services and too many machines. He told his readers Americans now "envied the Britisher" his smaller number of services and the higher quality of work turned out by British laundries–making a virtue out of what had previously been read as British industrial backwardness.[55]

By the same token, the meanings of laundry work's return to the home could be understood in a variety of ways. In a widely quoted passage from *Middletown: A Study in American Culture* (1929), Robert and Helen Lynd suggest that the choice of a washing machine over a laundry made little sense. They argued that the initial investment in a washing machine "tends to perpetuate a questionable institutional set-up–whereby many individual homes repeat common tasks day after day in isolated units." They painted a picture of a confused housewife manipulated into buying a washing machine by the "suave installment payment salesman," despite her own best interests.[56] Sales techniques certainly helped sell appliances, but women also had their own reasons for buying them or for saying "no," regardless of how suave the salesman.

Arguments that explain the housewife's choice of the washing machine over the laundry only in terms of the devaluation of women's work ignore the important component of class. The middle-class women who were most likely to be able to afford to choose between a laundry and a washing machine had long been beneficiaries of the monetary devaluation of working-class women's labor, in the form of domestic servants and cheap goods and services. If working women's wages had stayed at nineteenth-century levels, middle-class women might have been less eager to adopt washing machines. Thanks to the efforts of workers and reformers, the long-term pattern of devaluation of women's work lost some of its class basis, shifting the burden back onto middle-class women. By the postwar period, even a laundry worker might be able to afford her own washing machine.

CONCLUSION

Remembering the Laundry Industry

IN THE DECADES SINCE WORLD WAR II, LAUNDRIES DID NOT DISappear, but they did become invisible. Some hid behind storefronts advertising dry cleaning. Others, renamed as linen and diaper services, called at the back door and tucked their plants away in industrial districts and at the outskirts of cities. Today, a handful of shirt and family laundries can still be found in metropolises such as New York and London, but they are rare in the smaller towns and cities that once boasted a wide variety of services. Most middle-class Americans have forgotten how to fill out a laundry list, though nearly everyone knows how to operate a washing machine.

Contemporary public discussions have also abandoned the industry as a source of examples to demonstrate the evils and virtues of industrialization. It has been a long time since any leading reformer has declaimed the evils of laundry employment in a journal of opinion, and longer still since *Good Housekeeping* debated the relative merits of the washing machine and the commercial laundry. The industry remains nearly as invisible in the work of historians. Most seem more inclined to revisit the more obvious progenitors of our own world: steel and textiles, banks and trains. The industry's failure to triumph over its competitors, its small scale, and its identification with women and domesticity have all helped ensure its consignment to historical anonymity.

Why then bother to make laundries visible again? Part of the answer I offer is that they are good to think with (to borrow Claude Lévi-Strauss's phrase). The manifold functions of gender, the variable meanings of technology and technological knowledge, the economic and cultural world of the small-scale entrepreneur, all emerge in particularly sharp relief from this example. My point, therefore, in telling the industry's history, is neither to rescue it from historical oblivion nor to suggest a paradigm of industrialization centered on the laundry, but rather to find ways of looking anew at the bigger picture of industrialization.

What is visible and invisible in how we understand the past results not just from choice of subject matter but also from the boundaries historians necessarily draw around their subjects. Rethinking those parameters provides another means to reenvision what industrialization is about. Laundry owners, by definition the most local of business owners, engaged in a lively ongoing transatlantic conversation and an international exchange of technology, ignoring the constraints of nation and region when it suited their purposes. By the same token, the failure of borrowed ideas, particularly those about reform, to work outside the contexts in which they originated points to real differences in political cultures and social structures between the United States and Great Britain, and between the Northern and Southern states in America.

Similarly, separating workers, business owners, reformers, and consumers into their own discrete categories and histories certainly facilitates analytical neatness. But putting them back together in the same pages has its own rewards, revealing how complex, overlapping, and sometimes contradictory those roles can be. The laundry owners' slow realization that housewives acted as both consumers and competitors suggests greater porosity between the categories of producers and consumers that are often assumed. Boundaries between home and work, domestic and industrial, become equally permeable under close scrutiny.

Technology and technological change are indisputably a central element of the industrialization process. However, the processes of innovation and invention, so often central to histories of technology, represent only part of this story. As in many other industries, the basic technologies used in laundries came from borrowing and adapting devices employed in the home and in other industries. In this study I also posit the equal importance of another set of questions about technology: How did consumers, workers, and employers choose between existing technologies? How did technology and technological knowledge come to mean different things to different people? And what were the historical consequences of adopting technologies that function less than ideally in the hands of users?

Gender also has both obvious and not so obvious functions and meanings in this world. The existence of the industry itself depended on the mutual reinforcement of a gender system that claimed woman's place was in the home and an economic system that drove them into wage labor, dragging their irons behind them. However, laundry owners themselves also struggled with the complexities of gendered identities–questioning, for instance, what it meant to be a man doing laundry.

Telling the stories of small-scale industries as well as big business, failure and incompetence as well as triumph and ingenuity, helps to convey the breadth of ordinary people's experiences of industrialization. The vast majority of women never entered a steel mill, either as workers or as visitors. However, the changes industrialization brought to the ubiquitous problem of doing laundry almost inevitably touched their lives. For washerwomen, laundries provided competition. For the unemployed, they became a ready source of work. For the housewife, they were an alternative, albeit flawed, to the rigors of washday.

And while many men dreamed of becoming the next Henry Ford, most spent their days engaged in petty capitalism or wage labor. Like the other small businesses of which we know so little, laundries offered a way into the world of business requiring only a small amount of capital and a minimal education. While the corporate bureaucrat at Metropolitan Life or the Pennsylvania Railroad mastered the intricacies of paperwork and eyed the corporate ladder, the laundryman stuffed his receipts into an office safe and rolled up his sleeves if the washman or the boiler engineer failed to show up in the morning.

This book does not aim to call forth a laundrycentric paradigm of industrialization, nor does it exhaust the subject. Along with other aspects of industrialization, the laundry industry awaits other historians with new ideas about rendering visibility and invisibility, demonstrating connections and discontinuities, creating categories and disassembling them.

Notes

Introduction

1. For recent calls for a more transnational approach see George Fredrickson, "From Exceptionalism to Variability: Recent Developments in Cross-National Comparative History," *Journal of American History* 82 (September 1995): 587–604; Ian Tyrell, "American Exceptionalism in an Age of International History," *American Historical Review* 96 (October 1991): 1031–55; James E. Cronin, "Neither Exceptional nor Peculiar: Towards the Comparative Study of Labor in Advanced Society," *International Review of Social History* 38 (1993): 59–75.

2. See Nina Lerman, Arwen Mohun, and Ruth Oldenziel, "Versatile Tools: Gender Analysis and the History of Technology," *Technology and Culture* 38 (January 1997): 1–8.

3. The more general literature is vast. For an overview from a British history perspective, see Sonya O. Rose, "'Gender at Work': Sex, Class, and Industrial Capitalism," *History Workshop* 21 (1986): 113–31 and Sonya O. Rose, *Limited Livelihoods: Gender and Class in Nineteenth-Century England* (Berkeley, 1992), 1–21. On the United States, see Ava Baron, "Gender and Labor History: Learning from the Past, Looking to the Future," in *Work Engendered: Toward a New History of American Labor*, ed. Ava Baron (Ithaca, 1991). On the devaluation of women's work in a market economy, see Jeanne Boydston, *Home and Work: Housework, Wages, and the Ideology of Labor in the Early Republic* (New York, 1990).

4. Ulla Wikander, Alice Kessler-Harris, and Jane Lewis, eds., *Protecting Women: Labor Legislation in Europe, the United States, and Australia, 1880–1920* (Urbana, Ill., 1995); Susan Porter Benson, *Countercultures: Saleswomen, Managers, and Customers in American Department Stores, 1890–1940* (Urbana, Ill., 1988);

Angel Kwolek-Folland, *Engendering Business: Men and Women in the Corporate Office, 1870–1930* (Baltimore, 1994).

5. Patricia Malcolmson, *English Laundresses: A Social History, 1850–1930* (Urbana, Ill., 1986), provides an important starting point for understanding women who worked in laundries as well as the British reformers who tried to ameliorate the conditions under which women toiled.

6. Ruth Schwartz Cowan, *More Work for Mother: The Ironies of Household Technology from the Open Hearth to the Microwave* (New York, 1983), 105–7, 150. *More Work for Mother* is deservedly one of the most widely cited technological histories. However, Cowan's work has too often been taken not as a pioneering call for more research but rather as the first and last word on the enormously complex process she calls "the industrial revolution in the home."

7. For particularly influential examples, see Anthony Wallace, *Rockdale: The Growth of an American Village in the Early Industrial Revolution* (New York, 1980); Merritt Roe Smith, *Harper's Ferry Armory: The Challenge of Change* (Ithaca, 1977); and Judy McGaw, *Most Wonderful Machine: Mechanization and Social Change in Berkshire Paper Making, 1801–1885* (Princeton, 1987).

8. Benedict Anderson, *Imagined Communities: Reflections on the Origin and Spread of Nationalism* (London, 1983); Claude S. Fischer, *America Calling: A Social History of the Telephone* (Berkeley, 1992).

9. Maxine Berg, *Age of Manufactures: Industry, Innovation, and Work in Britain, 1700–1820* (Oxford, 1986); Raphael Samuel, "Workshop of the World: Steam Power and Hand Technology in Mid-Victorian Britain," *History Workshop* 3 (1977): 6–72; and Philip Scranton, *Figured Tapestry: Production, Markets, and Power in Philadelphia Textiles, 1885–1941* (Cambridge, Mass., 1989), 6.

1 / Technical and Cultural Origins

1. The most widely cited version of this story is in Fred De Armond, *The Laundry Industry* (New York, 1950), 7. For earlier versions, see "Our Twenty-Fifth Anniversary," *National Laundry Journal* 51 (1 January 1904): 12; "A History of the American Laundry Industry" (typescript, n.d.) Women's Bureau Records, Record Group 86, Box 105, National Archives, Washington, D.C.; "The Brief Story of the Progress of the Power Laundry Industry from Ancient Times" (typescript, 1928) Women's Bureau Records, Box 105, National Archives, Washington, D.C. See also Frederic W. Bradshaw, *Power Laundries: The Story of a Five Hundred Million Dollar Industry* (New York, 1926).

2. Caroline Davidson, *A Woman's Work Is Never Done: A History of Housework in the British Isles, 1650–1950* (London, 1982), 136–37. Henry Mayhew, observer of the London working class during the 1850s, disagreed with the contemporary assumption that single men invariably paid a woman to do their wash. Henry Mayhew, *London Labour and the London Poor*, vol. 2 (London, 1861), 190.

3. Ancliffe Prince, *The Craft of Laundering: A History of Fabric Cleaning* (London, 1970), 11. On dollies, see Pamela Sambrook, *Laundry Bygones* (Aylesbury, 1983), 7.

4. For a more extensive analysis, see Arwen Mohun, "Laundrymen Construct Their World," *Technology and Culture* 36 (January 1997): 97–120.

5. Siegfried Giedion, *Mechanization Takes Command: A Contribution to Anonymous History* (New York, 1948), 561.

6. Benjamin Butterworth, *The Growth of Industrial Art* (Washington, D.C., 1892), 84. The Shaker washing machine marketed to hotels and laundries between 1850 and 1870 is an exception in using a rubbing rather than agitating design. See *Improved Shaker Washing Machine* (Concord, N.H., 1859, 1877).

7. Sambrook, *Laundry Bygones*, 3–4; Linda Campbell Franklin, *Three Hundred Years of Housekeeping Collectibles* (Florence, Alabama, 1992), 64–132.

8. For examples, see: *Catalogue of the Troy Laundry Machinery Co.* (Troy, N.Y., 1901), 57, 69.

9. Davidson, *A Woman's Work*, 150; Susan Strasser, *Never Done: A History of American Housework* (New York, 1982), 105; Christina Walkey and Vanda Foster, *Crinolines and Crimping Irons: Victorian Clothes: How They Were Cleaned and Cared For* (London, 1978), 54.

10. Eliza Leslie, *The Housebook: Or, A Manual of Domestic Economy* (Philadelphia, 1840), 7.

11. Ibid., 8.

12. "First Report of the Commissioners on the State of Large Towns and Populous Districts with Minutes of Evidence, Appendix, and Index," No. 384 (1844) *Parliamentary Papers*, 260. See also Strasser, *Never Done*, 105.

13. Leslie, *The Housebook*, 9.

14. For more extensive description, see Davidson, *A Woman's Work*, 138–50; Strasser, *Never Done*, 105–9; and Walkey and Foster, *Crinolines and Crimping Irons*, 58–61.

15. Leslie, *The House Book*, 27; Catherine Beecher, *A Treatise on Domestic Economy* (1860; reprint, New York, 1977), 324.

16. Leslie, *The House Book*, 15; Sambrook, *Laundry Bygones*, 5.

17. Ruth Schwartz Cowan, *More Work for Mother: The Ironies of Household Technology from the Open Hearth to the Microwave* (New York, 1983), 98; Strasser, *Never Done*, 109; Walkey and Foster, *Crinolines and Crimping Irons*, 54.

18. Charles Sylvester (Engineer), *The Philosophy of Domestic Economy; as Exemplified in the Mode of Warming, Ventilating, Washing, Drying, and Cooking . . . Adopted in the Derbyshire General Infirmary, and More Recently, on a Greatly Extended Scale, in Several Other Public Buildings* (Nottingham, 1819), 11.

19. Ibid., v. For a more complete description of bleaching machinery, see Edward Baines, *A History of the Cotton Manufactures of Great Britain* (London, 1835), 250–51.

20. Sylvester, *Philosophy*, 28, 30, 11, 23, 9.

21. Stephen Glover, *History and Gazetteer of the County of Derby: Drawn Up from Actual Observation and from the Best Authorities*, vol. 1 (Derby, 1833), 145. To Americans, Jedidiah may be more well known as the master to whom Samuel Slater, pioneer of the American textile industry, was apprenticed. George S. White, *Memoir of Samuel Slater, the Father of American Manufactures* (1836; reprint, New York, 1967), 33.

22. R. S. Fitton and A. P. Wadsworth, *The Strutts and the Arkwrights, 1758–1830: A Study of the Early Factory System* (Manchester, 1958), 173.

23. Strutt's circle of correspondents on technical and scientific matters included members of the Birmingham Lunar Society, notably Thomas Edgeworth and Erasmus Darwin, as well as Charles Babbage and Jeremy and Samuel Bentham (Samuel was a naval engineer and inventor). Strutt was also a fellow of the Royal Society. "Jedidiah Strutt," *Dictionary of National Biography*, vol. 19 (Oxford, 1967–1968), 65.

24. Fitton and Wadsworth, *The Strutts*, 186, 177; William Strutt to Jeremy Bentham, 4 October 1827, Bentham Correspondence, vol. 1, fol. 169, British Library. See also a patent filed for William Shotwell, April 21, 1803, #3035 in "Patents for Inventions," *Abridgements of Specifications Relating to Washing and Wringing Machines, 1691–1866*, vol. 89 (London, 1877) n.p.

25. Giedion, *Mechanization Takes Command*, 564.

26. Sylvester, *Philosophy*, 57, 60–62.

27. *Subject-Matter Index of Patents for Inventions Issued by the United States Patent Office from 1790–1873, Inclusive* (Washington, D.C., 1874), 844, 1640.

28. Sir John Dalrymple, *Advantages of Public Washing Houses, in or Near Great Towns, Conducted by Steam Instead of Fuel and the Hand* (London, 1802), n.p.

29. Ibid.

30. "Another Strike amongst the Washerwomen," *Pioneer* (14 June 1834): 407.

31. Henry Mayhew, *The Morning Chronicle Survey of Labour and the Poor: The Metropolitan Districts* (Horsham, 1981), 111, 114.

32. Carroll Pursell, *The Stationary Steam Engine in America: A Study in the Migration of a Technology* (Washington, D.C., 1969), 87.

33. This is consistent with overall patterns of industrialization. See Raphael Samuel, "Workshop of the World: Steam Power and Hand Technology in Mid-Victorian Britain," *History Workshop* 3 (1977): 6–72.

34. *Ladies Philadelphia Shopping Guide and Housekeeper's Companion* (Philadelphia, 1859), 37.

35. William Burns, *Female Life in New York City Illustrated with Forty-four Portraits from Real Life* (Philadelphia, 1859), 94.

36. Caroline H. Dall, *Women's Right to Labor: Or, Low Wages and Hard Work* (Boston, 1860), 172.

37. "J. T. King's Washing Apparatus, Patented October 21, 1851" (Baltimore, 1852), 11.

38. "Description and Philosophy of James T. King's Patent Washing and Drying Apparatus" (New York, 1855), 7.

39. Ibid., 28.

40. "J. T. King's Washing Apparatus," 11.

41. The founding dates of these companies are self-ascribed. See "Summerscales . . . established 1850" [advertisement], 3; "Harper Twelve Trees . . . established 1845," *Laundry Journal* (15 January 1890): 3, 4. While the large British firms still extant in the early twentieth century could date themselves back to the 1850s, the Civil War provides a watershed in the existence of their American counterparts. James King's enterprise does not seem to have survived through the 1860s, and the first firms in Troy, New York–the late-nineteenth-century

center of laundry machinery manufacture–began in that decade. Most notable among them was A. P. Adams, Laundry Machinery and Supplies, later the Troy Laundry Machinery Company (by 1900 the mostly likely claimant of "largest laundry machinery manufacturer in the world"). Arthur James Waite, *The City of Troy and Its Vicinity* (Troy, N.Y., 1886), 347.

42. "History and Development of Laundry Machinery, *Laundry Journal* (16 November 1891): 14–15; "Manlove and Alliot," *Laundry Journal* (15 August 1890): 6; Alexander Alliot, British patent numbers 10,070 (24 February 1844) and 13,490 (3 February 1851); Prince, *The Craft of Laundering,* 12.

43. Isabella Beeton, *Beeton's Book of Household Management* (London, 1861), 1008.

44. Francis Place, add. MS 27827, fols. 50–51, Francis Place Letters, British Library, cited in *The Autobiography of Francis Place, 1771–1854,* ed. Mary Thale (Cambridge, 1972), 51. Place might have taken special notice, as his mother, aided by Place's wife, supported herself as a washerwoman in her old age. Ibid., 99.

45. Bonamy Dobree, ed., *The Letters of Philip Dormer Stanhope, Fourth Earl of Chesterfield,* vol. 3 (London, 1932), 973. In his youth, William Strutt was advised by his father to read Chesterfield's letters and take them as a model. Fitton and Wadsworth, *The Strutts,* 144.

46. Richard Bushman, *The Refinement of America: Persons, Houses, Cities* (New York, 1992), 41–42, 71.

47. The phrase is from John Wesley, Sermon 93, "On Dress" (1740).

48. Henry D. Rack, *Reasonable Enthusiast: John Wesley and the Rise of Methodism* (London, 1989), 535. John told Henry Moore that he dreaded his brother Charles's visiting him, despite his affection for him, because Charles was sure to disturb his arrangement of books and papers. Ibid., 252.

49. Georges Vigarello, *Concepts of Cleanliness: Changing Attitudes in France since the Middle Ages,* trans. Jean Birrell (Cambridge, 1988), 9.

50. Richard L. Bushman and Claudia L. Bushman, "The Early History of Cleanliness in America," *Journal of American History* 74 (1988): 1214. Mrs. Drinker's story was first told in Harold Donald Eberlein, "When Society First Took a Bath," in *Sickness and Health in America: Readings in the History of Medicine and Public Health,* ed. Judith Walzer Leavitt and Ronald L. Numbers (Madison, 1978), 331–41.

51. Bushman, "History of Cleanliness," 1215.

52. Vigarello, *Concepts of Cleanliness*, 46, 60, 76.

53. Patricia Malcolmson, *English Laundresses: A Social History, 1850–1930* (Urbana, Ill., 1986), 66.

54. John Duffy, *The Sanitarians: A History of American Public Health* (Urbana, Ill., 1990), 20; Anthony S. Wohl, *Endangered Lives: Public Health in Victorian Britain* (Cambridge, Mass., 1983), 72, 88–89.

55. Eric E. Lampard, "The Urbanizing World," in *The Victorian City: Images and Realities*, vol. 1 (London, 1973), 4. The 1801 census revealed that one-fifth of America's population lived in cities and towns with ten thousand or more inhabitants. U.S. figures calculated from U.S. Department of Commerce, *Historical Statistics of the United States: Colonial Times to 1970*, part 1 (Washington, D.C., 1975), 12. In order to be consistent, I have used populations of ten thousand or more to define urban. In 1800, 3.5 percent of Americans lived in urban cities.

56. Donald Joseph Bogue, *The Population of the United States: Historical Trends and Future Projections* (New York, 1985), 121.

57. Roy Porter, "Cleaning up the Great Wen: Public Health in Eighteenth-Century London," in *Living and Dying in London*, ed. W. F. Bynum and Roy Porter (London, 1991), 70–71.

58. For instance, Edwin Chadwick's famous 1842 report on the health of towns in Britain had a substantial influence on the nascent American public health movement. Duffy, *The Sanitarians*, 66.

59. See Nancy Tomes, "The Private Side of Public Health: Sanitary Science, Domestic Hygiene, and the Germ Theory, 1870–1900," *Bulletin of the History of Medicine* 64 (winter 1990): 509–39.

60. This analysis is informed by Mary Douglas, *Purity and Danger: An Analysis of Concepts of Pollution and Taboo* (New York, 1966), 32. See also Norbert Elias, *The Civilizing Process*, vol. 1 of *The History of Manners* (New York, 1978), 158.

61. Alternatively, historian Phyllis Palmer makes a feminist psychoanalytic argument for the gendering of cleanliness, based on the work of Dorothy Dinnerstein, Jane Flax, and Nancy Chodorow. See Phyllis Palmer, *Domesticity and Dirt: Housewives and Domestic Servants in the United States, 1920–1945* (Philadelphia, 1989), 141–44.

62. Cf. Suellen Hoy, *Chasing Dirt: The American Pursuit of Cleanliness* (New York, 1995), 7–8.

63. Historian Suellen Hoy claims English travelers saw antebellum Americans as dirty. However, as she points out, the people they commented on were almost invariably rural, poor, and located on the frontier. Hoy, *Chasing Dirt,* 7–8. For a contemporary example, see Frances Trollope, *Domestic Manners of the Americans* (1839; reprint, New York, 1949), 52–53.

64. Davidson, *A Woman's Work,* 115.

65. Harriet Martineau, "How to Make the Home Unhealthy," *Harper's* 1 (October 1850): 619.

66. The most extensive discussion of this idea is in Alain Corbin, *The Foul and the Fragrant: Odor and the French Social Imagination* (Cambridge, Mass., 1986). See also Elias, *The Civilizing Process,* 189–91.

67. For an extended discussion of the effects of urban anonymity, see John F. Kasson, *Rudeness and Civility: Manners in Nineteenth-Century Urban America* (New York, 1990), 100–107.

68. Burns, *Female Life,* 93.

69. *Baths and Washhouses for the Labouring Classes* (London, 1845), 8.

70. The origin of this phrase is uncertain. The Oxford English Dictionary places its first use in the 1830s.

71. Mary Douglas, *Purity and Danger: An Analysis of Concepts of Pollution and Taboo* (New York, 1966), 32.

72. Some Church of England divines gave sermons on sanitation because they had been explicitly instructed to do so as a way of blunting the effects of the 1847 cholera epidemic. E. R. Norman, *Church and Society in England, 1770–1970: A Historical Study* (Oxford, 1976), 132. For an example of a directive, see C. J. Blomfield, "A Pastoral Letter to the Clergy of the Diocese of London" (London, 1847).

73. Lucy Cameron, *Cleanliness Is Next to Godliness* (London, 1823), 15.

74. Parenthetical citations herein refer to pages in G. B. Dickson, "On Cleanliness" (Edinburgh, 1852), 1.

75. *Wash [An Address on Cleanliness]* (London, 1847), n.p.

76. Ibid.

77. Norbert Elias suggests that a critical part of the change in personal habits is the internalization of standards. He notes that in the modern period, habits are

condemned less so in relation to others: whether they are "troublesome or embarrassing to others, or because they betray a 'lack of respect.'" Instead, habits of moral behavior are "associated with embarrassment, fear, shame, or guilt, even when one is alone." Elias, *The Civilizing Process,* 150.

78. *Routledge's Manual of Etiquette* (London, 1860), 49.

79. Ibid., 12.

80. Emily Thornwell, *The Lady's Guide to Perfect Gentility* (Philadelphia, 1865), 14.

81. Horatio Alger, *Ragged Dick* (1866; reprint, New York, 1962), 57.

82. Stephen Reynolds, *A Poor Man's House* (Oxford, 1982), 57.

83. Robert Roberts, *The Classic Slum: Salford Life in the First Quarter of the Century* (Manchester, 1971), 22.

84. Ibid., 17.

85. Mrs. H. W. Beecher, *Motherly Talks with Young Housekeepers* (New York, 1873), 266. The term *slattern* originally meant "kitchen maid." Its shorter version is *slut.* See Palmer, *Domesticity and Dirt,* 140.

86. Nelson Manfred Blake, *Water for the Cities: A History of the Urban Water Supply Problem in the United States* (Syracuse, N.Y., 1956), 63–77. On Britain, see Davidson, *A Woman's Work,* 21–28.

87. Bernard Rudden, *The New River: A Legal History* (Oxford, 1984), 131.

88. Wohl, *Endangered Lives,* 69; Strasser, *Never Done,* 92–94; Davidson, *A Woman's Work,* 30.

89. George Sims, *How the Poor Live* (London, 1883), 39.

90. Davidson, *A Woman's Work,* 7; Strasser, *Never Done,* 86.

91. For examples, see *The Laundry Trade Directory and Launderer's Handbook with Diary* (London, 1908), iv; *The Laundry Journal Diary* (London, 1894), 5; C. A. Royce, *The Steam Laundry and Its Methods* (Chicago, 1894), 33–34.

92. *The Laundry Maid: Her Duties, and How to Perform Them* (London, 1877), 78, 80.

93. "From Exchanges on the Laundry Strike," *National Laundry Journal* 49 (15 May 1903): 21.

94. Malcolmson, *English Laundresses,* 129.

95. Stanley Buder, *Pullman: An Experiment in Industrial Order and Community Planning, 1880–1930* (New York, 1967), 18.

96. See *The Theory and Practice of Warming and Ventilating Public Buildings, Dwelling Houses, and Conservatories . . . by an Engineer* (London, 1824), 99, 103, 296; Commission for Promoting the Establishment of Baths and Washhouses, *Baths and Washhouses for the Labouring Classes* (London, 1845, 1852).

97. Walkey and Foster, *Crinolines and Crimping Irons,* 127.

98. Ibid., 128–29.

99. Anzia Yezierska, "Soap and Water" in *Hungry Hearts* (Boston, 1920), 163.

100. Ibid., 167.

2 / The Business of Laundry

1. Eric B. Dobson, *Hartonclean, 1884–1984* (Newcastle, 1983), 2–5. Joicey was also the son of a mechanical engineer who was a partner in a Newcastle engine works. See "James Joicey," *Dictionary of National Biography, 1931–1940* (Oxford, 1949), 493.

2. American Steam Laundry Company, *Minute Book,* 1689 1/1, 15 March 1884; 6 May 1884; 22 July 1884, Bird's Laundry Collection, Tyne and Wear Archives, Newcastle-upon-Tyne.

3. American Steam Laundry Company, *Minute Book,* 2 December 1884; 3 February 1885; 10 February 1885.

4. Dobson, *Hartonclean,* 5.

5. American Steam Laundry Company, *Minute Book,* 6 July 1885.

6. Armstrong built a better reputation on this venture, continuing that business until the Depression, the devaluation of the British pound, and increased import duties forced his company to sell British-made machinery. See "Service–Their Pleasure," *The Power Laundry* (4 May 1935): 657.

7. For an overview, see Dobson, *Hartonclean.*

8. Ancliffe Prince, *The Craft of Laundering: A History of Fabric Cleaning* (London, 1970), 18; Patricia Malcolmson, *English Laundresses: A Social History, 1850–1930* (Urbana, Ill., 1986), 127.

9. See Troy Laundry Machinery Company, Ltd., *Laundry Guide* (Chicago, 1911) frontispiece; Troy Laundry Machinery Company, Ltd., *Illustrated Catalogue*

(Chicago, 1906); Adams Laundry Machinery Company, *Adams Laundry Machinery Company, Manufacturers of High Grade Laundry Machinery* (Troy, N.Y., 1914). British trade literature also featured American machines in their descriptions of prescribed laundry practice. See, for example, the Troy products in Arthur Morris, *The Modern Laundry: Its Construction, Equipment, and Management* (London, 1909), 86.

10. *The Modern Laundry Guide: A Collection of the Best Articles Published in the National Laundry Journal* (Chicago, 1905), 8.

11. See, for instance, C. A. Royce, *The Steam Laundry and Its Methods: Essays Read at the Laundrymen's National Conventions* (Chicago, 1894), frontispiece.

12. U.S. Department of Commerce, Bureau of the Census, "Statistics of Steam Laundries" in *Manufactures: 1909* (Washington, D.C., 1910), 3, 1. Industry representatives noted that this was an undercount because census takers sometimes counted multiple laundries under the same ownership as one establishment. They probably also missed a number of small laundries that used a limited amount of machinery and kept a low profile, working with strictly cash business to avoid or underestimate taxes. See notes on methodology, ibid., 1.

13. "Mrs. Bosanquet on West London Laundries," *Laundry Journal* (23 March 1907): 2; "Supplement to the Annual Report of the Chief Inspector of Factories and Workshops for the Year 1905: Return of Persons Employed in 1904 in Workshops and Laundries," Cd. 3323 (1907), *Parliamentary Papers*, 403.

14. U.S. Census, "Statistics of Steam Laundries," 4. In comparison, the 1900 census counted 120,603 female cotton mill operatives and 43,497 tobacco workers and cigar makers–two of the largest employers of women in factory settings. U.S. Department of Commerce, Bureau of the Census, *Special Reports: Occupations at the Twelfth Census* (Washington, D.C., 1904), vol. 189; U.S. Department of Commerce, Bureau of the Census, *Special Reports: Statistics of Women at Work* (Washington, D.C., 1907), 56.

15. "Mrs. Bosanquet on West London Laundries," 3. "Supplement to Annual Report," Cd. 3323 (1907) *Parliamentary Papers*, 403.

16. U.S. Census, "Statistics of Steam Laundries," 3, 11, 7; U.S. Senate, *Report on Woman and Child Wage-Earners in the United States*, vol. 12, *Employment of Women in Laundries*, 61st Cong., 1911, 2d Sess., S. Doc. 645, 10, 9.

17. Ibid.

18. *The Industries of Cleveland* (Cleveland, 1888), 173.

19. *Men of Ohio in Nineteen Hundred* (Cleveland, 1901), 145.

20. Dobson, *Hartonclean*, 2.

21. H. Llewellyn Smith, ed., *London Industries II*, vol. 5 of Smith, ed., *The New Survey of London Life and Labour* (London, 1930–1935), 373.

22. New York State Department of Labor, *Working Conditions in New York Steam Laundries* (Albany, 1912), 11; U.S. Senate, *Employment of Women in Laundries*, 13–14.

23. C. Cadogan Rothery and H. O. Edmonds, *The Modern Laundry: Its Construction, Equipment and Management* (London, 1909), vol. 2, 8. See also David G. Thomas and Brenda J. Sowan, *The Walton Lodge Sanitary Laundry* (London, 1977).

24. U.S. Census, "Statistics of Steam Laundries," 6.

25. *Laundry Journal* (2 January 1899): 7.

26. "Partnership," *Laundry Journal* (15 August 1887): 21; "Ladies' Laundries," *Laundry Journal* (1 February 1899): 9.

27. American Institute of Laundering, "Leadership through the Years, 1883–1958: American Institute of Laundering 75th Anniversary," (booklet, 1958) in William Frew Long Papers, Western Reserve Historical Society, Cleveland, Ohio.

28. Ibid.

29. "Changed Ownership: The Woodbridge Laundry of Chicago under New Management," *National Laundry Journal* 57 (15 March 1907): 48.

30. Cecil Roberts, *The First Fifty Years of "Millbay"* (Plymouth, 1948), 13–14.

31. Social historians have only begun to discover these groups. See David Monod, "Culture without Class: Canada's Retailers and the Problem of Group Identification, 1890–1940," *Journal of Social History* 23 (spring 1995): 540. For related discussions, see J. A. McKenna and Richard G. Roger, "Control by Coercion: Employers' Associations and the Establishment of Industrial Order in the Building Industry of England and Wales, 1860–1914," *Business History Review* 59 (summer 1985): 203–31; R. Bean, "Employers Associations in the Port of Liverpool 1890–1914," *International Review of Social History* 21 (1976): 358–82.

32. The differences between these competing journals and the factions that they represented were extremely complex and changed over time. There were also smaller competing or specialized journals which had shorter lifespans, like the British journals *Laundry Institution and Engineer*, *Laundry*, and *Laundry Age*.

33. "Notes," *Laundry Journal* (6 April 1888): 5.

34. Fred De Armond, *The Laundry Industry* (New York, 1950), 9, 10. See also photograph of those first delegates in "First Launderers' Convention Ever Held in the World, LNA Chicago, October 1, 1883," *National Laundry Journal* 52 (1 September 1904): 20.

35. "A Laundry Trade Association," *Laundry Journal* (23 January 1886): 5.

36. "The New Association," *Laundry Journal* (23 February 1886): 3.

37. "Decision against the Laundrymen," *American Laundry Journal* 17 (November 1898): 15.

38. See Tony Freyer, *Regulating Big Business: Antitrust in Great Britain and America, 1880–1990* (Cambridge, 1992).

39. American trade journals regularly reported on both regional and national conventions. See, for example, "The New Jersey Convention," *American Laundry Journal* 30 (August 1911): 11; For Britain, see "Acton and West London Laundry Proprietor's Association 4th Annual Excursion," *Laundry Journal* (1 September 1886): 7; "Harrogate: The Second National Laundry Congress," *Laundry Journal* (31 May 1919): 193–205; "The American Convention," *The Power Laundry* (24 July 1920): 643.

40. *American Laundry Journal* 30 (July 1911): 7.

41. Tera W. Hunter, *To 'Joy My Freedom: Southern Black Women's Lives and Labor after the Civil War* (Cambridge, Mass., 1997), 208

42. See, for example, collected legal articles in *The Modern Laundry Guide*, 389–413. For Britain, see *The Laundry Trade Directory and Launderers' Handbook with Diary* (London, 1908), 40–42. *Laundry Journal* carried a regular column on legal matters. See, for example, "Legal and Criminal," *Laundry Journal* (20 April 1886): 5.

43. This is a topic that needs much further exploration. For an initial interpretation see Mark Carnes, "Iron John in the Gilded Age," *American Heritage* 5 (1993): 37–45; Clifford Putney, "Service over Secrecy: How Lodge-Style Fraternalism Yielded Popularity to Men's Service Clubs," *Journal of Popular Culture* 1 (spring 1993): 179–90.

44. For a detailed exploration of this theme see Arwen Mohun, "Laundrymen Construct Their World," *Technology and Culture* 38 (January 1997): 97–120.

45. "Tell These Men About It," (advertisement) *Laundry Age* 4 (1 October 1924): 181.

46. "Salvaging the Small Laundries," *The Power Laundry* (1 February 1935): 155.

47. "Report on the Work of H. M. Women Inspectors of Factories," Cd. 223 (1900), *Parliamentary Papers*, 379.

48. *Catalogue of the Troy Laundry Machinery Company* (Troy, N.Y., 1901).

49. See, for example, *Laundry Journal* (12 December 1885): 10; *Laundry Journal* (15 January 1890): 17.

50. Dolph was one of the "laundry machinery pioneers" who originally provided machinery to the collar factories in Troy, New York. The *National Laundry Journal* was long published out of Troy, although Dolph moved his company to Cincinnati before the turn of the century.

51. *The Laundry Journal Diary* (London, 1894), 5.

52. *The Laundry Trade Directory*, inside cover, 7.

53. *Laundry engineer* in the United States more often referred to the person who ran the power plant in the laundry. See for example "A Glimpse at the Licensing of Engineers," *National Laundry Journal* 53 (1 January 1905): 40.

54. *W. Summerscales and Sons, Ltd.: Makers of High-Class Laundry Machinery* (Keighly, 1897), 13.

55. *The Laundry Trade Directory*, 36.

56. *Catalogue of the Troy Laundry Machinery Company*, 69.

57. "She Will Make a Success of It," *National Laundry Journal* 44 (1 October 1902): 19.

58. "Employee" to the New York Women's Trade Union League, cited in Leslie Woodcock Tentler, *Wage-Earning Women: Industrial Work and Family Life in the United States, 1900–1930* (New York, 1979), 45.

59. Paul C. Siu, *The Chinese Laundryman: A Study in Social Isolation* (New York, 1987).

60. *American Laundry Journal* 30 (January 1912): 30.

61. Sucheng Chan, *This Bitter-Sweet Soil: The Chinese in California Agriculture, 1860–1910* (Berkeley, 1986), table 5.

62. Renquiu Yu, *To Save China, to Save Ourselves: The Chinese Hand Laundry Alliance of New York* (Philadelphia, 1992); Siu, *The Chinese Laundryman.*

63. Lillian Ruth Mathews, *Women in Trade Unions in San Francisco* (Berkeley, 1913), 35; Anti-Jap Laundry League, *Pacific Coast Convention of the Anti-Jap Laundry League* (San Francisco, 1908).

64. On the difficulties of using census figures for women's employment, see David M. Katzman, *Seven Days a Week: Women and Domestic Service in Industrializing America* (New York, 1978), 298–314.

65. Reports of the Immigration Commission, "Immigrants in Cities," vol. 1, *Summaries of Seven Cities* (Washington, D.C., 1911): 94–95, table 50, cited in Susan A. Glenn, *Daughters of the Shtetl: Life and Labor in the Immigrant Generation* (Ithaca, 1990), 72.

66. *Laundry Trade Directory*, 332.

67. "Laundries and Sanitation," *Laundry Journal* (15 March 1899): 17.

68. *Laundry Journal* (1 October 1891): 9.

3 / Inside the Laundry

1. See U.S. Department of Labor, Bureau of Labor Statistics, *Employment of Women in Power Laundries in Milwaukee* (Washington, D.C., 1913), 9; "Is Washing a Lost Art?" *Work and Leisure* 18 (November 1893): 289; Josephine Goldmark, *Fatigue and Efficiency: A Study in Industry* (New York, 1912), 489; Arthur Morris, *The Modern Laundry: Its Construction, Equipment, and Management* (London, 1909), v.

2. Elizabeth Beardsley Butler, *Women and the Trades: Pittsburgh, 1907–1908* (1909; reprint, Pittsburgh, 1984), 161.

3. Butler, *Women and the Trades*, 164.

4. "Report on the Work of H. M. Women Inspectors of Factories," Cd. 223 (1900), *Parliamentary Papers*, 381.

5. For ads featuring secondhand and reconditioned machinery, see the monthly classified pages of the *American Laundry Journal* and the *National Laundry Journal*. Newspapers in large cities sometimes also carried ads, see for instance the London *Times*, 3 November 1917, p. 14, col. 6.

6. On Bird's Laundry, see chap. 2. For a sense of the machine trade see the list of international dealers on the title page of *Catalogue of the Troy Laundry Machinery Company*, 12th ed. (Chicago, 1901); ads for American machinery (e.g., A. P. Adams) [advertisement for Troy Laundry Machinery] *Laundry Journal* (12

December 1885): 12; (Troy Laundry Machinery Company) *Laundry Journal* (1 September 1886): 10; *The Laundry Journal Diary* (London, 1894), inside covers; "Laundry Machinery Exports," and "American vs. Foreign Machinery," in *The Modern Laundry Guide: A Collection of the Best Articles Published in the National Laundry Journal* (Chicago, 1905), 325–28; "Laundry Machinery Exports for Five Years," *National Laundry Journal* 59 (15 June 1908): 48. Occupational medicine expert Thomas Oliver blamed American machinery for the increase in accidents in British laundries, see Thomas Oliver, ed., *The Dangerous Trades: The Historical, Social, and Legal Aspects of Industrial Occupations as Affecting Health by a Number of Experts* (London, 1902), 665.

7. Cecil Roberts, *The First Fifty Years at "Millbay"* (Plymouth, 1948), 30.

8. *The Laundry Trade Directory and Laundry Handbook with Diary* (London, 1908), 17; *Laundry Journal* (15 April 1887): 11; *The Modern Laundry Guide,* 16–17.

9. Mary Drake McFeely, *Lady Inspectors: The Campaign for a Better Workplace, 1893–1921* (Oxford, 1988), 39.

10. For autobiographies, see Flora Thompson, *Lark Rise to Candleford* (London, 1984); Alice Linton, *Not Expecting Miracles* (London, 1982); Kathleen Woodward, *Jipping Street: Childhood in a London Slum* (London, 1928). American narratives, such as Cynthia Woods, "You Have to Fight for Freedom," in *Rank and File: Personal Histories by Working-Class Organizers,* ed. Alice Lynd and Staughton Lynd (New York, 1988), describe the later interwar period.

11. For examples, see U.S. Senate, *Report on Condition of Women and Child Wage-Earners*: vol. 12, *Employment of Women in Laundries* (Washington, D.C., 1911), 61st Cong., 1911, 2d Sess., S. Doc. 645, 25–29; "Report on the Work of H. M. Women Inspectors," Cd. 223 (1900), *Parliamentary Papers,* 377–80; "Mrs. Bosanquet on West London Laundries," *Laundry Journal* (23 March 1907): 3; American Steam Laundry Company, *Minutes of the Director's Meetings,* 11 May 1906, Tyne and Wear Archives, Newcastle-upon-Tyne.

12. "The Steam Laundry: Its Practical Management," *Laundry Journal* (1 March 1899): 17.

13. Butler, *Women and the Trades,* 162.

14. New York Department of Labor, "A Study of Hygenic Conditions in Steam Laundries and Their Effect upon the Health of Workers," *Special Bulletin No. 130* (Albany, 1924), 65.

15. U.S. Senate, *Employment of Women in Laundries,* 12; "Annual Report of the

Chief Inspector of Factories and Workshops," C. 8561 (1897), *Parliamentary Papers*, 57. For effects of the Factory Acts on facilities, see *Laundry Trade Directory*, 40.

16. Oliver, ed., *Dangerous Trades*, 670.

17. "Lost Her Whole Scalp," *National Laundry Journal* 59 (1 May 1908): 56. "What to Wear to Work," *National Laundry Journal* 44 (15 July 1900): 25.

18. *Laundry Wrinkles for the House and Factory, with Which Is Incorporated the Laundry Manual, etc.* (London, n.d.), iii.

19. C. A. Royce, *The Steam Laundry and Its Methods* (Chicago, 1894), 9–10.

20. U.S. Senate, *Employment of Women in Laundries*, 19; *American Laundry Journal* 39 (August 1920) 23; "Should Clothes Be Taken from Houses Where There Is Infectious Disease?" *National Laundry Journal* 59 (15 April 1908): 26.

21. Royce, *The Steam Laundry*, 16, 18; "Vanmen have Sex Appeal," *Laundry Journal* (April 1935): 385.

22. *American Laundry Journal* 31 (July 1912): 17.

23. "Legal and Criminal," *Laundry Journal* (23 January 1886): 5; "A Very Dishonest Woman," *National Laundry Journal* 44 (1 July 1900): 48; Royce, *The Steam Laundry*, 10; Butler, *Women and the Trades*, 188.

24. Butler, *Women and the Trades*, 190–91.

25. Troy Laundry Machinery Company, Ltd., *Laundry Guide* (Chicago, 1911), 65; Royce, *The Steam Laundry*, 21–22; "A Good Marking System," *National Laundry Journal* 51 (15 January 1904): 20; *The Modern Laundry Guide*, 125.

26. *American Laundry Journal* 31 (December 1912): 19.

27. *American Laundry Journal* 32 (May 1913): 21. See also "Family Washing Problem," *National Laundry Journal* 49 (1 April 1903): 29.

28. *American Laundry Journal* 31 (July 1912): 7.

29. *The Laundry Trade Directory*, 53.

30. "The Fabric Adulteration Evil and the Remedy," *American Laundry Journal* 31 (July 1912): 7, 17.

31. Christina Walkey and Vanda Foster, *Crinolines and Crimping Irons: Victorian Clothes: How They Were Cleaned and Cared For* (London, 1978), 135–37. For a more detailed description of establishments that specialized in dyeing and

dry cleaning see Butler, *Women and the Trades,* 204–6. The British laundries Millbay and Bird's both experimented with dry cleaning during this period. See Cecil Roberts, *The First Fifty Years of "Millbay,"* 19; "Blanchissage, Nettoyage, et Teinture" (Bird's Laundry publicity booklet, ca. 1904) n.p., Bird's Laundry Collection, Tyne and Wear Archives, Newcastle-upon-Tyne.

32. U.S. Department of Labor, *Employment of Women in Power Laundries,* 35.

33. *Laundry Management: A Handbook for Use in Private and Public Laundries . . . by the Editor of "The Laundry Journal"* (London, 1889), 105.

34. Charles Booth, *Life and Labour of the People of London,* 2d ser., vol. 8 (New York, 1896), 255; W. Summerscales and Sons, Ltd., *Laundry Machinery* (London, 1897), 96.

35. "The Modern Steam Laundry," *Scientific American* 63 (14 January 1893): 23; *American Laundry Journal* 31 (February 1913): 27; "A New Class of Detergents," *The Power Laundry* (28 May 1932): 647. See also Troy Laundry Machine Company, Ltd., *Laundry Guide,* 13–14 for an extended discussion of prescribed practices for laundry chemistry. This guide also provides essentially the same recipe for soap as the 1893 *Scientific American* article.

36. U.S. Senate, *Employment of Women in Laundries,* 19; *The Modern Laundry Guide,* 153–54; Royce, *The Steam Laundry,* 40–41, 50–61.

37. "The Indespensible Part of the Wash," *National Laundry Journal* 44 (15 September 1900): 25; "Good and Bad Soaps," *National Laundry Journal* 44 (15 July 1900): 8.

38. U.S. Senate, *Employment of Women in Laundries,* 19.

39. U.S. Department of Labor, *Women in Power Laundries,* 34; New York Department of Labor, "A Study of Hygienic Conditions in Steam Laundries," 18–26; "Report on the Work of H. M. Women Inspectors of Factories," Cd. 223 (1900), *Parliamentary Papers,* 385.

40. Butler, *Women and the Trades,* 168.

41. *Catalogue of the Troy Laundry Machinery Company,* 12th ed. (Chicago, 1901), 34–35; *W. Summerscales and Sons, Ltd.: Makers of Modern High-Class Laundry Machinery* (Nottingham, 1897), 14. An overhead drive is also shown in the New York laundry described in *Scientific American* 63 (14 January 1893): 24. The overhead drive seems to have been phased out in the United States first.

42. New York Department of Labor, "Study of Hygenic Conditions in Steam

Laundries," 26; "Her Arm Wrenched Off," *National Laundry Journal* 44 (15 July 1900): 5.

43. "Laundry Labour," *Laundry Record* (2 February 1920): 161.

44. U.S. Senate, *Employment of Women in Laundries,* 19–20.

45. Rheta Childe Dorr, *A Woman of Fifty* (New York, 1924), 179.

46. Pamela Sambrook, *Laundry Bygones* (Aylesbury, 1983), 13; W. Summerscales and Sons, *Illustrated Catalogue* (Nottingham, 1895), 40.

47. "Leading off for That $100," *National Laundry Journal* 48 (1 July 1902): 20.

48. Katherine Anthony, *Mothers Who Must Earn* (New York, 1914), 185.

49. *Catalogue of the Troy Laundry Machinery Company,* 12th ed., 68.

50. [Advertisement for British Henrici Laundry Machinery Company], *Laundry Journal* (2 January 1889): 14–15.

51. *Laundry Trade Directory and Launderers' Handbook with Diary* (London, 1908), 19; "The Making and Use of Raw Starch," *National Laundry Journal* 44 (1 January 1900): 17.

52. Walkey and Foster, *Crinolines and Crimping Irons,* 64.

53. U.S. Senate, *Employment of Women in Laundries,* 20.

54. For a good description of the drying closet see: *Laundry Institution and Engineer* (15 October 1910): 6.

55. "Automatic Conveyor Drying Room," *National Laundry Journal* 44 (1 August 1900): 29; Butler, *Women and the Trades,* 176.

56. See advertisement for American Laundry Machinery Company dryer in *American Laundry Journal* 31 (December 1912): 14.

57. U.S. Senate, *Employment of Women in Laundries,* 13.

58. Patricia Malcolmson, *English Laundresses: A Social History* (Urbana, Ill., 1986), 143.

59. Sambrook, *Laundry Bygones,* 16–17.

60. W. Summerscales and Sons, *Catalogue* (1895), 44. The Board of the American Steam Laundry Company authorized the purchase of a set of gas irons in 1894. See American Steam Laundry Company, *Minute Book* (1 April 1894), Tyne and Wear Archives, Newcastle-upon-Tyne.

61. Thomas Oliver, ed., *Diseases of Occupation from the Legislative, Social, and Medical Points of View* (London, 1908), 389.

62. U.S. Senate, *Employment of Women in Laundries,* 31.

63. Linton, *Not Expecting Miracles,* 48.

64. See, for instance, "Number 5 Collar and Cuff Ironer" in *Catalogue of the Troy Laundry Machinery Company,* 12th ed., which the company claimed had been on the market for nearly a quarter of a century and in all those years had always been considered "the best collar and cuff ironer that was ever built" (72–73). The machine was displaced after about ten years. For a very early example of a specialized ironing machine for collars, see: M. Mahoney, *Catalogue and Price List* (Troy, N.Y., 1883), 22. This device was a small cylinder turned by hand with a press containing a hot slug of metal that clamped down on it. These devices were probably originally developed to be used in the Troy collar factories. For a discussion of the devices in use, see C. A. Royce, *The Steam Laundry,* 219.

65. *The Laundry Trade Directory and Handbook,* 28; U.S. Department of Labor, *Employment of Women in Power Laundries,* 77.

66. See U.S. Department of Labor, *Employment of Women in Power Laundries.* The ill effects of treadle machines was of special interest to the authors of this study. For illustrations of the wide variety of machines available after 1900, see Troy Laundry Machinery Company, *Illustrated Catalogue* (Chicago, 1907), 113–37.

67. Sue Ainslie Clark and Edith Wyatt, *Making Both Ends Meet: The Income and Outlay of New York Working Girls* (New York, 1911), 184.

68. Butler, *Women and the Trades,* 181.

69. *American Laundry Journal* 31 (December 1912): 17.

70. Butler, *Women and the Trades,* 187–88.

71. *American Laundry Journal* 31 (December 1912): 29.

4 / Women Workers and the Laundry Industry

1. Illinois Bureau of Labor Statistics, *Report of the Bureau of Labor Statistics of Illinois, 1906* (Springfield, 1908), 245.

2. See Alba M. Edwards, *Sixteenth Census of the United States: 1940, Population: Comparative Occupation Statistics for the United States, 1870–1940* (Washington, D.C., 1943).

3. See Dorothy Sue Cobble, "'Drawing the Line': The Construction of a Gendered Work Force in the Food Service Industry," in *Work Engendered: Towards a New History of American Labor*, ed. Ava Baron (Ithaca, 1991), 217; Patricia Cooper, *Once a Cigar Maker: Men, Women, and Work Culture in American Cigar Factories, 1900–1919* (Urbana, Ill., 1987).

4. Patricia Malcolmson, *English Laundresses: A Social History, 1850–1930* (Urbana, Ill., 1986), 16–17.

5. "Report of Principal Lady Inspector," C. 223 (1899) *Parliamentary Papers*, 258; "Royal Commission on Labour: Employment of Women in Laundries," C. 6894 (1893–94) *Parliamentary Papers*, 21; Thomas Oliver, ed., *Dangerous Trades: The Historical, Social, and Legal Aspects of Industrial Occupations as Affecting Health by a Number of Experts* (London, 1902), 667.

6. U.S. Senate, *Report on the Condition of Woman and Child Wage-Earners in the United States*, vol. 12, *Employment of Women in Laundries* (Washington, D.C., 1911), 13.

7. Charles Booth, *Life and Labour of the People of London*, 2d ser., vol. 8 (New York, 1896), 255.

8. Patricia Malcolmson, *English Laundresses: A Social History, 1850–1930* (Urbana, Ill., 1986), 11; Leonore Davidoff, "The Employment of Married Women in England, 1850–1950," (M.A. Thesis, London School of Economics, 1956), 216.

9. Booth, *Life and Labor*, 2d ser., vol. 8, 267; Katherine Anthony, *Mothers Who Must Earn* (New York, 1914), 106; U.S. Senate, *Employment of Women in Laundries*, 12–13; Malcolmson, *English Laundresses*, 12.

10. U.S. census data is not helpful in estimating the number of hand laundries, however a 1911 survey of 2,500 laundries in Chicago, New York, Brooklyn, Philadelphia, and Rockford, Illinois, claimed that only one-sixth of those laundries were "motor laundries." Chicago was the only city in which power laundries outnumbered hand laundries. U.S. Senate, *Employment of Women in Laundries*, 9.

11. Lack of American interest can also be explained by the fact that they began researching the industry later, after hand laundries seemed, as one report put it, part of the "procession of vanishing things." U.S. Department of Labor, Bureau of Labor Statistics, *Employment of Women in Power Laundries in Milwaukee* (Washington, D.C., 1913), 9. On the other hand, at least two prominent charities ran large hand laundries to train and give employment to needy women.

In New York, the Charity Organization Society operated the Park Avenue Laundry. See Box 139, Charity Service Society Archives, Columbia University Rare Books and Manuscripts Library, New York. See also Susan J. Kleinberg, *The Shadow of the Mills: Working-Class Families in Pittsburgh* (Pittsburgh, 1989), 247.

12. "Royal Commission on Labour," C. 6894 (1893–94) *Parliamentary Papers,* 33.

13. Ibid., 22.

14. "Legal and Criminal," *Laundry Journal* (15 August 1887): 11.

15. "The Laundry Act," *Laundry Journal* (1 August 1887): 5.

16. Sue Ainslie Clark and Edith Wyatt, *Making Both Ends Meet: The Income and Outlay of New York Working Girls* (New York, 1911), 196. Alice Henry, *The Trade Union Woman* (New York, 1915), 186; George R. Sims, *Off the Track in London* (London, 1911), 6, 41; Booth, *Life and Labour,* 1st ser., vol. 5, 90–91, 151–52.

17. For more on cultural images of the traditional laundress, see Caroline Davidson, *A Woman's Work Is Never Done: A History of Housework in the British Isles, 1650–1950* (London, 1982), 56, 136–37; Malcolmson, *English Laundresses,* 5–6, 104–5. Barbara Littlewood and Linda Mahood argue that these qualities were used to identify prostitutes in nineteenth-century Britain. See Littlewood and Mahood, "Prostitutes, Magdalenes, and Wayward Girls: Dangerous Sexualities of Working Class Women in Victorian Scotland," in *Gender and History* 3 (summer 1991): 160–75. They also argue that when these women were institutionalized, they were forced to do laundry work as a "cleansing ritual."

18. "Royal Commission on Labour," C. 6894 (1893–94) *Parliamentary Papers,* 17.

19. *Laundry Journal* (23 March 1907): 1.

20. Sims, *Off the Track in London,* 28–30.

21. U.S. Senate, *Employment of Women in Laundries,* 50.

22. One British study claimed that nearly half of all female workers in British power laundries were hand ironers. This seems unlikely, but I have found no other figures to contradict it. "Women in Industry: Report of the War Cabinet Committee on Women in Industry," Cmd. 135 (1919), *Parliamentary Papers,* 54. In the United States, out of the sample of 539 laundry workers discussed in the 1911 Senate study, only 59 were ironers. U.S. Senate, *Employment of Women in Laundries,* 38. Neither British nor American census compilations make a detailed enough distinction to be of use.

23. U.S. Senate, *Employment of Women in Laundries,* 17–18. The high wages of ironers also undermined initial efforts to get laundries included under the 1909 Trade Board Act as a sweated industry. "The Laundry Trade Board Act: Statement of Evidence," Lab 2/TB 102, p. 2, Public Record Office, Kew Gardens. See Chapter 8 for further discussion of the Trade Boards.

24. For a more extensive discussion of the definition of "skill" in this context, see Arwen Mohun, "Why Mrs. Harrison Never Learned to Iron: Gender, Skill, and Mechanization in the American Steam Laundry Industry," *Gender and History* 8 (August 1996): 231–51.

25. *Report on the Laundry of the Charity Organization of New York* (typescript, 1907), n.p., Box 139, Charity Service Society Archives, Columbia University Rare Books and Manuscripts Library, New York City.

26. Elizabeth Beardsley Butler, *Women and the Trades: Pittsburgh, 1907–1908* (1909; reprint, Pittsburgh, 1984), 163.

27. By *apprenticeship* I mean an institution less strict and legally binding than the formal indentured apprenticeship that characterized preindustrial artisanal culture–an institution already breaking down in both Britain and the United States during this period, even in the trades.

28. "Royal Commission on Labour," C. 6894 (1893–94) *Parliamentary Papers,* 19.

29. Anthony, *Mothers Who Must Earn,* 76.

30. Though several weeks or months does not seem like a long training period compared to the multiyear training periods common in male-dominated work, in the context of women's employment, it constituted an unusual and substantial investment in time and effort.

31. Anthony, *Mothers Who Must Earn,* 76.

32. "Royal Commission on Labor," C. 6894 (1893–94) *Parliamentry Papers,* 18.

33. The Women's Industrial Council, *Annual Report,* 1905–1906 (London, 1906), 17.

34. Ibid.

35. Malcolmson, *English Laundresses,* 41.

36. U.S. Senate, *Employment of Women in Laundries,* 12.

37. Clark and Wyatt, *Making Both Ends Meet,* 189.

38. Ibid., 182.

39. Ibid., 188, 189.

40. Butler, *Women and the Trades,* 178.

41. Cecil Roberts, *The First Fifty Years of "Millbay"* (Plymouth, 1948), 33.

42. U.S. Department of Commerce, Bureau of the Census, *Fourteenth Census of the United States,* vol. 10, *Manufactures: 1919, Reports for Selected Industries,* "Power Laundries" (Washington, D.C., 1923), 1026.

43. "Interview with Mr. R. W. Barber, Spier and Pond's Laundry," *Laundry Journal* (15 July 1899).

44. Thomas Oliver, ed., *Dangerous Trades,* 665.

45. The British Laundress's closest American equivalent would have been the black washerwoman of Southern mythology (and, in some cases, Southern reality). See C. G. Woodson, "The Negro Washerwoman, a Vanishing Figure," *Journal of Negro History* 15 (July 1930): 269–77. Woodson describes this figure as a "towering personage" and attributes her disappearance to the growth of the steam laundry industry (269–70). There are no lamentations for her passing in the pages of the trade journals.

46. "Mrs. Bosanquet on West London Laundries," *Laundry Journal* (23 March 1907): 3.

47. U.S. Census categories did not distinguish between washerwomen, hand laundry, and steam laundry workers before 1900.

48. New York Department of Health, *The Cost of Clean Clothes in Terms of Health* (Albany, 1918), 35. Other state bureaus did similar studies; see also Massachusetts Department of Labor, "Wages of Women in Laundries in Massachusetts," *Bulletin of the Minimum Wage Division* 5 (Boston, 1914).

49. See U.S. Senate, *Employment of Women in Laundries,* 39, passim.

50. "It Was the Soap," *National Laundry Journal* 52 (1 November 1904): 18. In her study of women workers in Chicago, Joanne Meyerowitz notes that "in laundries and kitchens, black women often received one dollar less per week than white women." Joanne J. Meyerowitz, *Women Adrift: Independent Wage Earners in Chicago, 1880–1930* (Chicago, 1988), 36.

51. Butler, *Women and the Trades,* 182.

52. Wyatt and Clark, *Making Both Ends Meet,* 195; U.S. Senate *Employment of Women in Laundries,* 71–75.

53. U.S. Senate, *Employment of Women in Laundries,* 72.

54. "Ladies in Laundries," *Work and Leisure: The Englishwoman's Advertiser, Reporter, and Gazette* 14 (March 1889): 74.

55. Ibid., 75.

56. Anthony, *Mothers Who Must Earn,* 85–86.

57. Butler, *Women and the Trades,* 183–84.

58. "Coming of Age Party," *Laundry Journal* (22 June 1907): 15.

59. "Latchford Steam Laundry Social" and "Hygenic Laundry Social," *Laundry Journal* (19 January 1907): 3.

60. "Laundries Will Not Have Outing on Account of Convention," *Laundry Journal* (15 August 1902): 5.

61. Even in the wash house and steam plant, the manageress was legally responsible for accidents and for this reason the *Laundry Journal* suggested that they should at least understand the working principles of an engine and perhaps even be able to fire it up in an emergency. "The Steam Laundry: Its Practical Management–Part 3: The Washhouse," *Laundry Journal* (1 March 1899): 17.

62. One of the ways laundry owners attempted to attract middle-class women into laundry management was through articles in women's magazines. See, for example, "Ladies and Laundries," 113–16; Mrs. Philipps, *A Dictionary of Employments Open to Women* (London, 1898), 84–85; G. E. Mitton, ed., *The Englishwoman's Year Book and Directory* (London, 1915), 113–14; "Laundry Work as a Profession for Educated Women," *The Woman Leader* (31 March 1922): 68.

63. "Interview with Miss Prince of the New Era Laundry," *Laundry Journal* (1 August 1890): 12.

64. Dorothy Richardson, *The Long Day: The Story of a New York Working Girl as Told by Herself* (New York, 1905), 237–38.

65. "The Ladies," *Laundry Journal* (15 March 1899): 11.

66. "Women for the Trade Board," *Laundry Record* (2 August 1920): 527.

67. "Mr. Hugh Trenchard on Lady Managers," *Laundry Journal* (15 June 1899): 15.

68. "Waste Girls," *Laundry Journal* (1 September 1887): 2.

69. See Martha Vicinus, *Independent Women: Work and Community for Single Women, 1850–1920* (Chicago, 1985).

70. "Cottage Laundries," *Laundry Journal* (April 1890): 18.

5 / Unionization and Regulation, 1888–1914

1. See Robin Miller Jacoby, *The British and American Women's Trade Union Leagues, 1890–1925* (Brooklyn, 1994), 9.

2. See Candice Dalrymple, *Sexual Distinctions in the Law: Early Maximum Hour Decisions of the United States Supreme Court, 1905–1917* (New York, 1987).

3. Norbert Soldon, *Women in British Trade Unions, 1874–1976* (Totowa, N.J., 1978), 54; *The Laundry Journal Diary* (London, 1894), 19; "Amendment to the Factory and Workshop Act of 1901," *Laundry Journal* (25 May 1907): 1, 3; "Act up for Revision," *Laundry Journal* (8 June 1907): 1, 6–7; "Discussion of Bill at Laundry Association Meeting," *Laundry Journal* (6 July 1907): 3.

4. See, for example "Not Fundamental at All," *National Laundry Journal* 49 (1 June 1903): 1; "Labor Unions vs. Laundries," *National Laundry Journal* 48 (15 October 1902): 8; "Doom of the Closed Shop Agreement," *National Laundry Journal* 52 (15 July 1904): 8.

5. "Meeting of Wandsworth Laundry Workers," *Women's Union Journal* 14 (15 February 1889): 15; Women's Trade Union League, "Fifteenth Annual Report," (1889) *Women's Trade Union League Papers* (microfilm; Wakefield, West Yorkshire, 1975), reel 1.

6. Phillipa Levine, *Victorian Feminism, 1850–1900* (Tallahassee, Fla., 1987), 108.

7. "The Society of Laundresses," *Women's Union Journal* 14 (15 October 1889): 86; Mary Drake McFeely, *Lady Inspectors: The Campaign for a Better Workplace, 1893–1921* (Oxford, 1988), 9; Soldon, *Women in British Trade Unions*, 35, 36; Barbara Drake, *Women in Trade Unions* (London, 1984), 28.

8. "Meeting of the Wandsworth Laundry Workers," *Women's Union Journal* 14 (15 February 1889): 15; "To the Editor," *Women's Union Journal* 15 (15 August 1890): 70; Women's Trade Union League, "Fifteenth Annual Report," 6.

9. Gertrude Tuckwell, "The More Obvious Defects in Our Factory Code," in *The Case for the Factory Acts*, ed. Beatrice Webb (London, 1901), 157. On the Women's Trade Union League and protective legislation, see: Rosemary Feurer, "The Meaning of 'Sisterhood': The British Women's Movement and Protective Labor Legislation, 1870–1900," *Victorian Studies* 31 (winter 1988): 234; Jacoby, *British and American Women's Trade Union Leagues*, 43; Soldon, *Women in British Trade Unions*, 35. By the 1880s, limited gender-specific regulation of hours and conditions of work was an accepted part of British public policy. The word-

ing of the Act's 1878 revisions inadvertently excluded laundries, only encompassing industries such as textile manufactures, producing goods for sale, or those such as dye works, substantially altering the goods upon which they worked.

10. "The Trades Union Congress, 1890," *Englishwoman's Review* (15 October 1890): 383, 385; see also "Correspondence," *Women's Union Journal* 15 (15 August 1890): 70.

11. Drake, *Women in Trade Unions,* 27.

12. *Times* (London), 15 June 1891.

13. Tuckwell, "The More Obvious Defects," 158; Drake, *Women in Trade Unions,* 28.

14. *Times* (London), 15 June 1891.

15. Ibid.

16. "The Laundresses' Agitation," *Women's Trade Union Review* (July 1891): 1. Others lobbied members of Parliament and passed out handbills. See Tuckwell, "The More Obvious Defects," 158.

17. *Punch* (20 June 1891), 290; Drake, *Women in Trade Unions,* 28.

18. [Advertisement], *Laundry Journal* (1 September 1890): 13; McFeely, *Lady Inspectors,* 5.

19. "Onward," *Laundry Journal* (15 October 1890): 5.

20. Patricia Malcolmson, *English Laundresses: A Social History, 1850–1930* (Urbana, Ill., 1986), 51. Malcolmson provides a detailed analysis of the debate between pro- and antiprotectionist feminists. For the broader context, see Levine, *Victorian Feminism,* 118–22.

21. *Laundry Journal* (1 May 1891): 13.

22. *Hansard's Parliamentary Debates,* 3rd series, 354 (19 June 1891), 928–47, 940.

23. Ibid., 936. As word spread, "the members of the Grand Committee on Trade were besieged by letters and petitions from convents and homes, clergymen and philanthropists, Anglicans and Roman Catholics." See Helen Bosanquet, Louise Creighton, and Beatrice Webb, "Law and the Laundry," *Nineteenth Century* 41 (1897): 225.

24. Tuckwell, "The More Obvious Defects," 158.

25. *Hansard's Parliamentary Debates,* 947.

26. "Laundries and Legislation," *Englishwoman's Review* (16 October 1893): 22–23; Drake, *Women in Trade Unions,* 28.

27. McFeely, *Lady Inspectors,* 36; "Royal Commission on Labour: Employment of Women in Laundries," C. 6894 (1893–94) *Parliamentary Papers,* 17–23.

28. Bosanquet et al., "Law and the Laundry," 226.

29. George McCleary, *Life in the Laundry* (London, 1902), 10–11.

30. Gertrude Tuckwell, "The Government Factory Bill of 1900," *Fortnightly Review,* new ser. 73 (1900): 972.

31. For example, see ibid., 976; Margaret McDonald, *Labour Laws for Women: Their Reason and Their Result* (London, 1900), 11–12; *The Dangerous Trades: The Historical, Social, and Legal Aspects of Industrial Occupations as Affecting Health by a Number of Experts,* ed. Thomas Oliver (London, 1902), 663–64. See also *Muller v. Oregon,* 208 U.S. 412, 421 (1908).

32. McFeely, *Lady Inspectors,* 15, 23, 26.

33. Ibid., 39–40; Rose Squire, *Thirty Years in the Public Service: An Industrial Retrospect* (London, 1927), 25–26; B. Leigh Hutchins and Amy Harrison, *A History of Factory Inspection* (New York, 1970), 195; Adelaide Anderson, *Women in the Factory: An Administrative Adventure, 1893–1921* (New York, 1922), 141–46.

34. Women's Trade Union League, "Thirty-First Annual Report" (1906) *Women's Trade Union League Papers,* reel 1, 14.

35. "Annual Report on the Work of H. M. Women Inspectors During 1897," C. 8965 (1898) *Parliamentary Papers.*

36. *The Workmen's Compensation Act: What It Means, and How to Use It* (London, 1898), 4.

37. Vivien Hart, *Bound by Our Constitution: Women, Workers, and the Minimum Wage* (Princeton, 1994), 21. See Chapter 7 for more on the Trade Boards.

38. Drake, *Women in Trade Unions,* 28; "Royal Commission on Labour," C. 6894 (1893–94), *Parliamentary Papers,* 23, 37; Malcolmson, *English Laundresses,* 116.

39. Levine, *Victorian Feminism,* 108; Norbert Soldon, "British Women and Trade Unionism: Opportunities Made and Missed," in *The World of Women's Trade Unionism: Comparative Historical Essays,* ed. Norbert Soldon (London, 1985), 15.

40. See "13th Report of the Chief Labour Correspondent of the Board of Trade on

Trade Unions for 1900 (with Retrospective to 1892)," Cd. 773 (1901), *Parliamentary Papers*; "17th Report of the Chief Labor Correspondent," Cd. 777 (1905), *Parliamentary Papers.*

41. Soldon, *Women in British Trade Unions,* 57–58.

42. Malcolmson, *English Laundresses,* 116; Soldon, *Women in British Trade Unions,* 69.

43. Malcolmson, *English Laundresses,* 116, 119.

44. McDonald, *Labour Laws for Women,* 7–8. For an overview of female union membership during the period 1876–1918, see Drake, *Women in Trade Unions,* Table 1.

45. This is Patricia Malcolmson's primary explanation. See Malcolmson, *English Laundresses,* 116.

46. As an early historian of British women's unions observed: "Women were badly paid because they were unorganized and unorganized because they were badly paid." See Drake, *Women in Trade Unions,* 57.

47. On the collar laundresses, see Carole Turbin, *Working Women of Collar City: Gender, Class, and Community in Troy, 1864–86* (Urbana, Ill., 1992); Philip S. Foner, *Women and the American Labor Movement from Colonial Times to the Eve of World War I* (New York, 1979), 155. See also Ruth Delzell, *Early History of Women Trade Unionists of America* (Chicago, 1919), 15. The Knights laundryworkers and laundry employees' assemblies included Brooklyn #6346 (1886–1891), Albany #7075 (1886–1890), Cleveland #529-b (1894–1896), and Seattle #2632-b (1890–1891), cited in Jonathan Garlock, *Guide to the Local Assemblies of the Knights of Labor* (Westport, Conn., 1983). See also John B. Andrews and W. D. P. Bliss, *History of Women in Trade Unions* (New York, 1974), 130.

48. See Tera W. Hunter, *To 'Joy My Freedom: Southern Black Women's Lives and Labors after the Civil War* (Cambridge, Mass., 1997).

49. Foner, *Women and the American Labor Movement,* 244.

50. Philip Foner has suggested that this may also explain why Samuel Gompers was initially reluctant to grant a charter to the new union. See ibid., 244–45.

51. [Editorial Cartoon], *New York Times,* 7 January 1912.

52. Douglas Pearson Hoover, "Women in Nineteenth Century Pullman" (M.A. Thesis, University of Arizona, 1988), 29–30; Stanley Buder, *Pullman: An Experiment in Industrial Order and Community Planning, 1880–1930* (New York, 1967),

171–74. For more on the Pullman laundries, see Mrs. Duane Doty, *The Town of Pullman* (Pullman, Ill., 1893) 126–33; Pullman Papers, esp. Payroll Records, Newberry Library Archives, Chicago.

53. *Chicago Tribune,* 24 and 27 April, 1903.

54. *Chicago Record-Herald,* 4 May 1903.

55. "It Was a Black Friday," *National Laundry Journal* 49 (15 May 1903): 1.

56. *Chicago Chronicle,* 27 April 1903.

57. *Chicago Record-Herald,* 4 May 1903.

58. T. E. Wilson, identified as both the president and the business manager of the local laundry union, told the *Chicago Tribune* that the union had been started the previous August by organizers from the AFL: *Chicago Tribune* (2 May 1903). Emily Barrows, in "Trade Union Organization among Women" (Ph.D. dissertation, University of Chicago, 1925), 157, states that the union was originally a local women's union with no AFL affiliation. I have found no other evidence to support her assertion.

59. "Peace Where There Is No Peace," *National Laundry Journal* 49 (1 June 1903): 1.

60. *Chicago Tribune,* 14 and 17 May 1903.

61. Chicago *American,* 4 May, 1903.

62. *Chicago Record-Herald,* 9 May, 1903.

63. *National Laundry Journal* 49 (15 May 1903): 22.

64. *Chicago Tribune,* 6 May 1903.

65. *Chicago Record-Herald,* 4 May 1903.

66. "It Was a Black Friday," *National Laundry Journal* 49 (15 May 1903): 26.

67. *Chicago Inter-Ocean,* 2 June 1903. For more on Dowie, see Rolvix Harlan, *John Alexander Dowie and the Christian Apostolic Church* (Evansville, Ill., 1906).

68. *National Laundry Journal* 49 (15 June 1903): 2.

69. *Chicago Inter-Ocean,* 2 June 1903.

70. *Chicago Record-Herald,* 9 June 1903; "The Long Strike Ended," *National Laundry Journal* 49 (1 June 1903): 17.

71. "The Long Strike Ended," *National Laundry Journal* 49 (1 June 1903): 1. When Emily Barrows investigated trade union organization amongst women in Chicago in the 1920s, she found that there was no laundry union in the city since it had collapsed in 1915. Attempts to reorganize workers during World War I had proved unsuccessful. Barrows, "Trade Union Organization," 57, 61, 64.

72. Lillian Ruth Mathews, *Women in Trade Unions in San Francisco* (Berkeley, 1913), 11.

73. Ibid., 12.

74. Ibid., 14.

75. "Editorial," *National Laundry Journal* 57 (15 June 1907): 2.

76. Mathews, *Women in Trade Unions,* 35; Robert M. Robinson, "A History of the Teamsters in the San Francisco Bay Area, 1850–1950" (Ph.D. dissertation, University of California, Berkeley, 1951), 137; Robert Edward Lee Knight, *Industrial Relations in the San Francisco Bay Area, 1900–1918* (Berkeley, 1960), 191. See Chapter 3 for a more extensive discussion of the Anti-Jap Laundry League.

77. Mathews, *Women in Trade Unions,* 20.

78. Ibid., 20–21, 38

79. Ibid, 38.

80. Andrews and Bliss, *History of Women in Trade Unions,* 139; U.S. Department of Commerce, Bureau of the Census, *Manufactures: 1909,* "Census of Steam Laundries" (Washington, D.C., 1910), 4.

81. *New York Times,* 2 January 1912. The Women's Trade Union League stated after the fact that for women workers, the strike was about hours, not wages. See Mary E. Dreier, "To Wash or Not to Wash: Ay, There's the Rub," *Life and Labor* 2 (March 1912): 68.

82. *New York Times,* 2 January 1912.

83. *New York Times,* 5 January 1912.

84. Inspired by a successful strike of collar starchers in Troy, the New York WTUL had brought in an organizer from San Francisco who visited nearly one hundred laundries in New York City but found little enthusiasm for union activity. In 1906 the WTUL laundry committee disbanded, declaring that its effort had been "untimely" and that the laundresses were "under too high pressure and too timid to accept action." See Nancy Schrom Dye, *As Equals and as Sisters: Feminism, the Labor Movement, and the Women's Trade Union League*

of New York (Columbia, Mo., 1980), 64–65.

85. *New York Times,* 6 January 1912.

86. *New York Times,* 15 January and 17 January 1912.

87. Dye, *As Equals,* 102.

88. Ibid.

89. See Judith Baer, *The Chains of Protection: The Judicial Response to Women's Labor Legislation* (Westport, Conn., 1978); Dalrymple, *Sexual Distinctions in the Law*; Nancy S. Erickson, "*Muller v. Oregon* Reconsidered: The Origin of a Sex-Based Doctrine of Liberty of Contract," in *Labor History* 30 (spring 1989): 228–50; Susan Lehrer, *Origins of Protective Legislation for Women, 1905–1925* (Albany, 1987); Alice Kessler-Harris, "Law and a Living: The Gendered Content of 'Free Labor,'" in *Gender, Class, Race, and Reform in the Progressive Era,* ed. Noralee Frankel and Nancy S. Dye (Lexington, Ky., 1991), 87–109; Nancy Woloch, *Muller v. Oregon: A Brief History with Documents* (Boston, 1996). For a comparative perspective, see Ulla Wikander, Alice Kessler-Harris, and Jane Lewis, eds., *Protecting Women: Labor Legislation in Europe, the United States, and Australia, 1880–1920* (Urbana, Ill., 1995).

90. Woloch, *Muller v. Oregon,* 18–19.

91. Josephine Goldmark, *Impatient Crusader: Florence Kelley's Life Story* (Urbana, Ill., 1953), 144–45.

92. Dalrymple, *Sexual Distinctions in the Law,* 132.

93. Ibid.

94. Goldmark, *Impatient Crusader,* 150.

95. Ibid., 154. Emphasis in original.

96. Ibid., 156.

97. Cited in Woloch, *Muller v. Oregon,* 117, 126, 128.

98. Ibid., 127, 129.

99. For a broader perspective on this issue, see Robert Gray, "The Languages of Factory Reform in Britain, c. 1830–1860," in *The Historical Meanings of Work,* ed. Patrick Joyce (Cambridge, 1987), 145.

100. Goldmark, *Impatient Crusader,* 157.

101. *Muller v. Oregon,* 208 U.S. 412, 421, 422 (1908).

102. "Editorial," *National Laundry Journal* 59 (15 March 1908): 2.

103. Agnes Nestor, *Woman's Labor Leader* (Rockford, Ill., 1954), 94; "The Eight Hour Fight at Springfield," *Union Labor Advocate* 10 (July 1909): 22, 23.

104. Goldmark, *Impatient Crusader*, 143.

105. Nestor, *Woman's Labor Leader*, 90–91.

106. Chicago Women's Trade Union League, *Biennial Report, 1911–1913* (Chicago, 1913), 19.

6 / Production and Consumption during the Interwar Period

1. For multigenerational laundries, see Papers of Byrd's Laundry, Tyne and Wear Archives, Newcastle-upon-Tyne, and Millbay Laundry, Devon County Record Office, Plymouth.

2. "How Shall We Revive Starched Collars?" *Laundry Age* 1 (1 March 1921): 48.

3. "The Laundry Bill," *The Power Laundry* (8 February 1919): 45.

4. "Why Laundries Lose Patrons," *Laundry Age* 1 (1 March 1921): 7.

5. "A Billion Dollar Business by 1930," *Laundry Age* 4 (1 September 1924): 2.

6. For an exploration of these concepts, see Philip Scranton, "Diversity in Diversity: Flexible Production and American Industrialization, 1880–1930," *Business History Review* 65 (spring 1991): 27–90.

7. "Housewives Crying for Laundry Service," *Laundry Age* 4 (1 September 1924): 99.

8. Ibid.

9. "Customer's Complaints: The Housewife and the Modern Laundry," *The Power Laundry* (20 August 1921): 723.

10. See, for instance, "The Housewife and the Modern Laundry," *The Power Laundry* (20 August 1921): 723; "The Customers' Point of View," *Laundry Age* 1 (1 May 1921): 18; "Housewife Fires Broadside at Laundry," *Laundry Age* 1 (1 July 1921): 26. For an analysis of home economists as technological mediators, see Carolyn Goldstein, "From Service to Sales: Home Economics in Light and Power, 1920–1940," *Technology and Culture* 37 (January 1997): 121–52.

11. "A Woman's View of Laundry Service," *The Power Laundry* (1 June 1935): 877.

12. "Just Look at That Shirt," *Laundry Age* 16 (April 1936): 14.

13. "Customer's Complaints," *The Power Laundry* (6 August 1921).

14. "Autocratic Laundry," *The Power Laundry* (4 May 1935): 691.

15. "Romance Dies beside the Wash Tubs," *Laundry Age* 16 (1 April 1936): 56.

16. "Handling the Family Wash," *Laundry Age* 1 (1 April 1921): 78.

17. "Getting the Public Eye," *Laundry Age* 16 (1 January 1936): 36. See Roland Marchand, *Advertising the American Dream: Making Way for Modernity, 1920–1940* (Berkeley, 1985), 341–47 for more on this strategy.

18. On radio programs, see "Radio Program in Jersey," *Laundry Age* 16 (1 February 1936): 77. For film advertising, see "Cornered! They See, Hear, and BUY!" *Laundry Age* 16 (1 January 1936): 22.

19. "Unusual Opening Planned for Community Laundry," *National Laundry Journal* 89 (15 April 1923): 68.

20. *Good Housekeeping* 87 (December 1928): 183.

21. "Laundry 'at Homes,'" *The Power Laundry* (3 September 1921): 754.

22. "Twenty-Five Years of Laundry Development," *The Power Laundry* 15 (4 May 1935): 648.

23. "Says Silas Kidd the Suds Philosopher: Ask Henry Ford," *Laundry Age* 16 (1 March 1936): 6.

24. *The Power Laundry* (11 January 1919): 3.

25. "Semi-finished Shirt Seen as Business Boon," *Laundry Age* 16 (1 January 1936): 26.

26. "Effects of the Great Social Changes," *The Power Laundry* (4 May 1935): 638. "Five Lucky Washdays," *National Laundry Journal* 92 (1 July 1924): 9.

27. "The Customer and the Laundry," *The Power Laundry* (12 January 1929): 31.

28. "Effects of the Great Social Changes," *The Power Laundry* (4 May 1935): 638.

29. "Which Laundry Service Does the Consumer Prefer?" *Laundry Age* 16 (1 May 1936): 48.

30. "Is Damp Wash on the Dole?" *Laundry Age* 16 (April 1936).

31. This transformation was most akin to the process going on among retailers. See Susan Strasser, *Satisfaction Guaranteed: The Making of the Mass Market*

(New York, 1989). In other respects, the analogy does not hold. The most important relationships small retailers had were those with suppliers and those with the competition from a growing number of chains. In contrast, large and small laundry owners often shared a common cause (although large laundries sometimes used trucks to poach on the territory of small town launderers). Laundry owners' most important relationships were with their customer-competitors.

32. See review of J. Ellis Barker, *Britain's Industrial Backwardness* in *Laundry Record* (May 1920): 311. For a more general discussion, see *Competitiveness and the State: Government and Business in Twentieth-Century Britain*, ed. Geoffrey Jones and Maurice Kirby (Manchester, 1991) and Michael Dintenfass, *Industrial Decline of Industrial Britain, 1870–1980* (London, 1992).

33. Charles was Cecil Roberts's father. See Cecil Roberts, *The First Fifty Years of "Millbay"* (Plymouth, 1948), 42. See also "American Methods," *Laundry Record* (1 November 1920): 887.

34. "The British Bundle," *Laundry Age* (1 January 1936): 99. "As Our English Cousin Sees Us," *National Laundry Journal* 89 (15 January 1923): 12.

35. Ibid.

36. J. J. Stark, "American Methods," *Laundry Record* (1 November 1920): 888.

37. "Effects of the Great Social Changes," *The Power Laundry* (4 May 1935): 638.

38. Roberts, *First Fifty Years*, 37.

39. "Laundry as a Hobby," *The Power Laundry* (12 November 1921): 995. See also "Why Some Laundrymen Fail in Business," *The Power Laundry* (19 April 1919): 139 (this article was apparently reprinted from the regional American laundry journal *The Pacific Laundryman*).

40. "Businessmen vs. Alleged Businessmen," *National Laundry Journal* 89 (15 January 1923): 56.

41. "Are Laundry Schools Worthwhile?" *National Laundry Journal* 89 (1 June 1923): 8.

42. "New York City Laundryowners Challenged," *Laundry Age* 4 (1 September 1924): 63. In the 1920s, there were several training programs in the United States. The oldest was at the Ohio Mechanics Institute, which ran a two-year course, and the Dunwoody Industrial Institute in Minneapolis also had a course. The public schools in New York, Detroit and Buffalo also ran programs in cooperation with local trade associations. See "Are Laundry Schools Worth-

while?" *National Laundry Journal* 89 (1 June 1923): 28. For British examples, see "Laundry Science," *Laundry Record* (January 1920): 25; "Twenty-Five Years of Laundry Development," *The Power Laundry* (4 May 1935): 648.

43. At least some of the students in these programs seem to have been the children of laundry owners. For the example of a British laundry owner's daughter who attended the American Institute of Laundering, see "Miss Hall Manages Plant," *Laundry Age* 16 (1 February 1936): 84.

44. "Thirty-Third National Convention," *National Laundry Journal* 76 (15 October 1916): 4.

45. "Federation," *Laundry Record* (2 February 1920): 93.

46. Neither the LNA nor the British organizations ever collected dues from a majority of business owners. Although trade associations had various strategies for a sliding scale of dues, owners of small laundries were less likely to join than owners and managers of large-scale concerns. In 1924, the LNA estimated that membership reflected only 25 percent of the total number of power laundries in the United States. "Anyone Can Inhale the Frangrance of a Rose," *National Laundry Journal* 92 (1 August 1924): 64.

47. On the United States, see L. E. Warford and Richard May, *Trade Association Activities* (Washington, D.C., 1923); C. Judkins and L. Marceron, *Selected Trade Associations of the United States* (Washington, D.C., 1936). The most closely analogous secondary literature is on association among retailers. Retailers and laundry owners had very different problems with regard to competition, which was reflected in the policies, organization, and ideology of trade associations. See Chris Hosgood, "A 'Brave and Daring Folk'? Shopkeepers and Trade Associations in Victorian and Edwardian England," *Journal of Social History* 26 (winter 1992): 285–308; and David Monod, "Culture without Class: Canada's Retailers and the Problem of Group Identification, 1890–1940," *Journal of Social History* 29 (spring 1995): 521–45.

48. "Building Better Business," *National Laundry Journal* 76 (15 July 1916): 2.

49. *The Power Laundry* (21 August 1920): 735; "Monopoly Frowned on Again," *Laundry Age* 16 (1 January 1936): 4.

50. Eric B. Dobson, *Hartonclean, 1884–1984* (Newcastle-upon-Tyne, 1983), 19; "Minutes of the Director's Meetings," 17 December 1906 and 6 February 1907, Bird's Laundry, Tyne and Wear Archives. The agreement entered into by Bird's lasted in various forms into the 1920s.

51. "No More Crank Handle or Whale Oil," *Laundry Age* 16 (1 January 1936): 24–25.

52. "What the Industry Needs Most," *Laundry Age* 1 (1 March 1921): 1.

53. "What the Laundry Industry Needs Most," *Laundry Age* 1 (1 April 1921): 32.

54. "What the LNA is Doing," *Laundry Age* 1 (1 March 1921): 18–19. See also, for example, "Adding New Pages to Your Book of Youth," *National Laundry Journal* 92 (1 August 1924): 15, which was reprinted in the August 23 issue of the *Saturday Evening Post.*

55. "Many Americans Visit England," *Laundry Age* 4 (1 September 1924): 92.

56. "Twenty Five Years of Laundry Development," *The Power Laundry* (4 May 1935): 648.

57. Sinclair Lewis, *Babbitt* (1922; reprint, New York, 1985), 87.

58. "What the LNA is Doing," *Laundry Age* 4 (1 March 1921): 18.

59. Ibid.

60. "Let's Stop Apologizing," *Laundry Age* 16 (1 April 1936): 2.

61. "Many Americans Visit England," *Laundry Age* 4 (1 September 1924): 92.

62. "The British Bundle," *Laundry Age* 16 (1 January 1936): 99.

63. "A Woman's Point of View," *The Power Laundry* (31 May 1919): 205. See also "Women for the Trade Board," *Laundry Record* (2 August 1920): 457.

64. "Twenty-Five Years of Laundry Development," *The Power Laundry* (4 May 1935): 648.

65. "Woman Launderer Invents Flexible Hydro-Divider," *Laundry Record* (27 July 1938): 127.

66. "A Merry Party at Grange-Over-Sands," *The Power Laundry* (5 March 1932): 273.

67. "Success Crowns the First Four States Convention," *National Laundry Journal* 92 (1 July 1924): 3.

7 / Laundries in Black and White

1. See Jacqueline Jones, *Labor of Love, Labor of Sorrow: Black Women, Work, and the Family from Slavery to the Present* (New York, 1985); James Grossman, *Land of Hope: Chicago, Black Southerners, and the Great Migration* (Chicago,

1989); Marvin E. Goodwin, *Black Migration in America from 1915 to 1960: An Uneasy Exodus* (Lewiston, N.Y., 1990).

2. Chicago Commission on Race Relations, *The Negro in Chicago: A Study of Race Relations and a Race Riot* (Chicago, 1923), 358.

3. Ibid.

4. Census takers during different decades were given varying directions about how to categorize workers. One of the consequences is a confusion between commercial and home laundresses, as well as between home laundresses and domestic servants who sometimes did laundry. See Joseph A. Hill, *Women in Gainful Occupations, 1870–1910*, Census Monograph IX (Washington, D.C., 1929), 112. Hill found a decrease in the number of domestic laundresses between 1910 and 1920 but suggested this might have been caused by a counting error. U.S. Department of Commerce, Bureau of the Sixteenth Census, *Comparative Occupation Statistics for the United States, 1870–1940* (Washington, D.C., 1943), 130. Bureau of the Fourteenth Census, *Occupations: Color or Race* (Washington, D.C., 1923), 358; Bureau of the Fifteenth Census, *Population* vol. 4, *Occupation by States* (Washington, D.C., 1933), 34.

5. Hill, *Women in Gainful Occupations*, 120.

6. Many of these women may have participated in agricultural work, which was not listed as one of the categories for previous employment.

7. Ethel L. Best and Ethel Erickson, *A Survey of Laundries and Their Women Workers in Twenty-three Cities* (Washington, D.C., 1930), 154.

8. Hill, *Women in Gainful Occupations*, 109.

9. U.S. Department of Commerce, Bureau of the Fourteenth Census, *Occupations: Males and Females in Selected Occupations* (Washington, D.C., 1920), 906.

10. Sylvia Woods, "You Have to Fight for Freedom," in *Rank and File: Personal Histories by Working-Class Organizers*, ed. Alice Lynd and Staughton Lynd (New York, 1988), 109–10.

11. Best and Erickson, *Survey*, 68.

12. Bertha M. Nienberg and Bertha Blair, "Factors Affecting Wages in Power Laundries," *Bulletin of the Women's Bureau* 14 (1936) 4, 34–35 (tables).

13. Best and Erickson, *Survey*, 61–85. Very detailed information can also be found in the individual schedules collected by investigators. See Women's Bureau Records, Record Group 86, National Archives, Washington, D.C.

14. Ibid., 69, 92.

15. Studies of relationships between black and white women workers are especially rare. Many of the studies that do exist look at these relations in the context of the South. See, esp., Dolores E. Janiewski, *Sisterhood Denied: Race, Gender, and Class in a New South Community* (Philadelphia, 1985); Mary Fredrickson "'I Know Which Side I'm on': Southern Women in the Labor Movement in the Twentieth Century," in *Women, Work, and Protest,* ed. Ruth Milkman (Boston, 1985), 156–80.

16. "How to Handle Negro Help in the Steam Laundry?" *National Laundry Journal* 62 (1 October 1909): 16.

17. Jane Filley, *Consider the Laundry Workers* (New York, 1937), 48, 24.

18. Woods, *Rank and File,* 112–13. Woods's story is inconsistent about how strictly jobs were racially divided. She claims there were "black jobs" and "white jobs" but also describes black women working white jobs for lower wages.

19. For a suggestive discussion of the culture of resistance among African American workers, see Robin D. G. Kelley, *Race Rebels: Culture, Politics, and the Black Working Class* (New York, 1994), 17–34.

20. See Chapter 8 for an in-depth discussion of Southern attitudes and their consequences.

21. For a description of the network of social workers, settlement house affiliates, clubwomen and others who were early supporters of the NAACP, the National Urban League, and organizations centered on women (such as the National League for the Protection of Colored Women) see Jesse Thomas Moore Jr., *A Search for Equality: The National Urban League, 1910–1961* (University Park, Penn., 1981), 40–41, 50–51; Dorothy C. Salem, "To Better Our World: Black Women in Organized Reform, 1890–1920," (Ph.D. dissertation, Kent State University, 1986).

22. Mary Van Kleeck, "New Standards for Negro Women in Industry," *Life and Labor* (June 1919): 134.

23. "Pilgrim's Progress in the Laundry," *Life and Labor* (April 1920): 118–19. The author also explains her surprise to find that the woman she describes as "Mammy" was the highest paid in the laundry because she could do any laundry job well.

24. In part, this strategy was consistent with the approach of many labor activists in this period. Frank Crosswaith, a founder of the National Urban League's Trade Union Committee for Organizing Negro Workers expressed his hope that the organization would "do for Negro Workers in New York City what the Women's Trade Union League does for women workers." *History of the Trade Union Committee* (n.p., ca. 1926) Box 8, National Urban League Industrial Relations Department, General Department File, National Urban League Records, Library of Congress, Washington, D.C.

8 / Bringing in the State

1. Cecil Roberts, *The First Fifty Years of "Millbay"* (Plymouth, 1948), 36.

2. Ibid., 44.

3. Vivien Hart, *Bound by Our Constitution: Women, Workers, and the Minimum Wage* (Princeton, 1994); Sheila Blackburn, "Ideology and Social Policy: The Origins of the Trade Board Act," *Historical Journal* 34 (March 1991): 43–64. For statistics on wages as a percentage of total costs, see for example: U.S. Department of Commerce, Bureau of the Census, *Fourteenth Census*, vol. 10, *Manufactures: 1919, Reports for Selected Industries, "Power Laundries"* (Washington, D.C., 1923), 1026.

4. For a similar interpretation, see Morton Keller, *Regulating a New Economy: Public Policy and Economic Change in America, 1900–1933* (Cambridge, 1990), 5–6.

5. The inclusion of trade associations in policymaking was characteristic not only of Hoover's associationalist state and predecessor organizations in America but also of British policymaking from at least the introduction of the Trade Board Acts in 1909. On the American pattern, see Brian Balogh, "Reorganizing the Organizational Synthesis: Federal Professional Relations in Modern America," *Studies in American Political Development* 5 (spring 1991): 119–71; Alan Dawley, *Struggles for Justice: Social Responsibility and the Liberal State* (Cambridge, Mass., 1991).

6. For a theoretical discussion of some of these issues see: Theda Skocpol, *Protecting Soldiers and Mothers: The Political Origins of Social Policy in the United States* (Cambridge, Mass., 1992); and Linda Gordon, "The New Feminist Scholarship on the Welfare State," in *Women, the State, and Welfare*, ed. Linda Gordon (Madison, 1990), 9–35.

7. While a single-industry study can only begin to speak to the vast literature on state regulation in this period, this comparison does point towards certain in-

terpretations. First, the strong similarities between both programs make it difficult to accept inclusive claims for American or British exceptionalism. See, in particular, arguments for the "triumph of conservatism" by followers of Gabriel Kolko in Colin Gordon, *New Deals: Business, Labor, and Politics in America, 1920–1935* (London, 1994); Ellis Hawley, *The New Deal and the Problem of Monopoly* (Princeton, 1966); Robert F. Himmelberg, *The Origins of the National Recovery Administration: Business, Government, and the Trade Association Issue, 1921–1933*, 2d ed. (New York, 1993). For a trenchant critique of the Kolko argument, see Keller, *Regulating a New Economy*, 4. Some of the literature on labor law is also relevant here. In particular, see the discussion on American exceptionalism and "reflective" versus "constitutive" theories of law in William E. Forbath, *Law and the Shaping of the American Labor Movement* (Cambridge, Mass., 1991), x, 2–11.

8. The NRA bore an even closer resemblance to the Whitley Council Scheme, proposed after World War I as an alternative to the Trade Boards. Industries were to be "voluntarily governed by a council composed of representatives of organized employers and organized workers . . . whose decisions should be binding upon all members of the organizations represented." Dorothy Sells, *The British Trade Board System* (London, 1923), 3. The Ministry of Labour found the laundry industry unsuitable for a Whitley Council, much to the disappointment of some. Decisions of a Whitley Council were not legally enforceable, however, whereas Trade Board decisions were. See "The Trade Boards Question," *The Power Laundry* (11 January 1919): 1 and "A Trade Board for the Laundry Industry," *The Power Laundry* (25 January 1919): 3.

9. The most suggestive work on the transatlantic culture of reform remains Morton Keller's brief essay, "Anglo-American Politics, 1890–1930," *Comparative Studies in Society and History* 22 (1980): 458–87. See also Kenneth O. Morgan, "The Future at Work: Anglo-American Progressivism 1890–1917," in *Contrast and Connection: Bicentennial Essays in Anglo-American History*, ed. H. C. Allen and Roger Thompson (London, 1976), 245–71 and C. L. Mowat, "Social Legislation in Britain and the United States in the Early Twentieth Century: A Problem in the History of Ideas," *Historical Studies* 7 (1969): 81–96.

10. Writing about the acts for an American audience of New Deal policymakers in 1939, British reformer Dorothy Sells characterized the broad intentions of the act as the "democratization" of income levels and of the wage-fixing process. Dorothy Sells, *British Wage Boards: A Study in Industrial Democracy* (Washington, D.C., 1939), 331. Most other historians have concluded the initial intention

was more narrowly pragmatic. Hart, *Bound by Our Constitution,* 4. Before the First World War, the act applied only to workers in the so-called sweated trades under the original act–chainmaking, tailoring, paper box making, machine lace making and net finishing–all characterized by "acute underpayment" according to the National Anti-sweating League. See Sells, *The British Trade Board System,* 2; Hart, *Bound by Our Constitution,* 17.

11. Sells, *The British Trade Board System,* 7.

12. "The Laundry Trade Board Act: Statement of Evidence," (typescript, 1914) 2, Lab 2/TB 102, Public Records Office, Kew Gardens (hereafter PRO).

13. "Report on the Proposed Extensions of the Trade Board Act to Laundry Workers," (January 1914), Lab 2/TB 102/1923, PRO.

14. "The Trade Boards Question," *The Power Laundry* (11 January 1919): 1; and "Can Organized Labour Assist Us?" *The Power Laundry* (8 February 1919): 41.

15. Sells, *The British Trade Board System,* 20.

16. "The Campaign against Laundries," *The Power Laundry* (6 September 1919): 356.

17. Mary Drake McFeely, *Lady Inspectors: The Campaign for a Better Workplace, 1893–1921* (Oxford, 1988), 150.

18. Patricia Malcolmson, *English Laundresses: A Social History, 1850–1930* (Urbana, Ill., 1986), 119–21.

19. Sells, *The British Wage Boards,* 334.

20. Also included were representatives of the National Union of General Workers, the Dockers' Union, the Workers' Union, the Amalgamated Society of Dyers, Bleachers, and Finishers, and the National Amalgamated Union of Labour. The meeting was presided over by W. G. Wardle, M.P. "Laundries Conference Held at the Ministry of Labour Tuesday February 5, 1919," (typescript, 1919) Lab 2/482/TB 10, PRO.

21. Sells, *The British Trade Board System,* 6.

22. Sells, *British Wage Boards,* 6.

23. Sells, *The British Trade Board System,* 165.

24. "Statement of Evidence," Lab 2/1923/TB 20, PRO.

25. Sells, *The British Trade Board System,* 89.

26. "Laundries Conference," Lab 2/482/TB 10, PRO.

27. "Slattery's Model Laundry–Minutes," Lab 2/1646/TB 62335, PRO; Albert Holloway to E. G. Wilson, 19 December 1928, Lab 2/1646/TB 50240, PRO. Holloway was a teenage boy who was asked by his employer to take on a variety of jobs previously done by adult men. For this work he was paid minimal wages.

28. "Minutes," (7 April 1937) Trade Board Meeting, Lab 11/TB 66, PRO.

29. Sells, *The British Trade Board System,* 153.

30. "Statement of Evidence," Lab 2/1923/TB 20, PRO.

31. See Blackburn, "Ideology and Social Policy" for a summary of various critiques of the acts. Malcolmson offers a seemingly contradictory assessment in *English Laundresses,* 100, 121.

32. Roberts, *First Fifty Years,* 37.

33. Hart, *Bound by Our Constitution,* 67.

34. Mary Elizabeth Pidgeon, "Women in the Economy of the United States of America: A Summary Report," *Women's Bureau Bulletin* 155 (1937), 101, 102–3; Irving Bernstein, *The Lean Years: A History of the American Worker, 1920–1933* (Baltimore, 1966), 225–26.

35. Hart, *Bound by Our Constitution,* 80.

36. Frances Perkins, *The Roosevelt I Knew* (New York, 1946), 55, 258. Perkins also drafted a trade board-based bill to replace the defunct NRA in 1936. See ibid., 254.

37. Vivien Hart, "Feminism and Bureaucracy: The Minimum Wage Experiment in the District of Columbia," *Journal of American Studies* 26 (April 1992): 1–22.

38. "New York Wage Board Gets into Action," *Laundry Age* 13 (1 August 1933): 14.

39. Edward Robb Ellis, *A Nation in Torment: The Great American Depression, 1929–1939* (New York, 1995), 340.

40. *National Industrial Recovery Act,* 13 June 1933, H.R. 5755, 73d Cong., 1st Sess. 1.

41. W. Conover to G. Larner, 8 June 1934, Box 3084, National Recovery Administration Records, Record Group 9, National Archives, Washington, D.C. (hereafter NRA Records).

42. Bertha M. Nienberg and Bertha Blair, "Factors Affecting Wages in Power Laundries," *Women's Bureau Bulletin* 143 (1936), 10; Laundryowner's National Association, "Information Requested on Industries," Box 3079, NRA Records.

43. Laundryowners' National Association, "Newsletter," (23 June 1933) Box 3084, NRA Records.

44. *Laundry Age* 13 (1 August 1933): 2. The president's Re-employment Agreement was an agreement between individual laundryowners and the president. The laundry code was supposed to be signed by the LNA on behalf of the entire industry. After being signed into law by the president, it would become binding for all members of the industry, whether they had directly agreed to it or not. In practice, the "local option" described below partially removed this requirement. In areas where the final NRA laundry code was never agreed to, laundryowners who had signed the agreement were supposed to continue to abide by it in order to be able to display their blue eagle, the symbol of NRA compliance.

45. Nienberg and Blair, *Factors*, 10; "Hearing," 578, Box 253, NRA Records.

46. W. W. Baker to Hugh Johnson, 30 June 1933, Box 3086, NRA Records.

47. Robert Weaver to Frances Perkins, 17 November 1933, in the Appendix to "Hearing," 194–95, Box 253, NRA Records; Clark Foreman to Frances Perkins, 28 November 1933, Box 3084, NRA Records. For more on Weaver and Foreman, see John Kirby, *The New Deal Era and Blacks: A Study of Black and White Race Thought, 1933–1945* (Urbana, Ill., 1971).

48. See the Appendix in "Hearing," 197–210, Box 253, NRA Records; "Analysis of Protests against Laundry Code," Box 3077, NRA.

49. M. J. Byrd to Rose Schneiderman, "Memorandum: The Negro Laundry Operative," Box 3086, NRA Records.

50. The AAUW was supposed to seek advice from the Bureau of Standards and the American Home Economics Association on setting standards for quality of service. The AAUW's principal agendas were to get laundries to use less caustic chemicals and to lose fewer items. "Hearing," 496–97, Box 253, NRA Records.

51. Ibid., 596.

52. Ibid., 515, 575, 578.

53. "Supplementary Brief on Wages," 28 November 1933, Box 3086, NRA Records.

54. "Back to Local Self-Regulation," *Laundry Age* 14 (1 March 1934): 2; "Local Option Governs Laundry Code," *Laundry Age* 14 (1 March 1934): 66; "Marking Time on the Code," *Laundry Age* 14 (1 April 1934): 14.

55. "The Why of the Local Option," *Laundry Age* 14 (1 May 1934): 90.

56. Mrs. M. A. Focht to A. D. Whiteside (n.d.), Box 3084, NRA Records.

57. "NRA Enters the Enforcement Stage," *Laundry Age* 14 (1 May 1934): 50, 52.

58. "Bulletin," *Laundry Age* 14 (June 1934): 3; "Feathers Fly," *Laundry Age* 14 (1 June 1934): 73.

59. "Lost: One Laundry Code. No Reward to Finder," *Laundry Age* 15 (1 April 1935): 87. State minimum wage laws were finally definitely upheld as constitutional by *West Coast Hotel Company v. Parrish,* in 1937.

60. "State Code Requested of New York Legislature," *Laundry Age* 14 (August 1934): 67; "Minimum Wage Set for New Hampshire Laundries," *Laundry Age* 14 (August 1934): 68; "The Plus Factor in Wages," *Laundry Age* 17 (1 February 1937): 2.

61. Mary Anderson to Gustav Peck, 14 December 1934, Box 3086, NRA Records; A. Howard Myers to Mary Anderson, 25 January 1935; Paul Hutchings to A. Howard Myers, 31 January 1935, Box 3084, NRA Records; John P. Davis to Myers, 25 January 1935, Box 3084, NRA Records. Davis was the executive secretary of the Joint Committee on National Recovery–an advisory group for African American rights in the New Deal. He asked Myers to make sure the Women's Bureau broke down all their data by race, sex, and occupation.

62. See Correspondence folder, Box 280, Women's Bureau Records, Record Group 86, National Archives, Washington D.C.

63. Nienberg and Blair, *Factors,* 4, 6–7, 12. *Laundry Age* also published a long article summarizing the bureau's findings: "Low Wages Not Caused by Low Prices," *Laundry Age* 17 (1 February 1937): 10–12.

9 / Compliance and Enforcement at the Other End of the Stick

1. The NRA's inability to enforce compliance has been widely remarked upon. See Colin Gordon, *New Deals: Business, Labor, and Politics in America, 1920–1935* (Cambridge, 1994), 191–92. No recent study discusses enforcement of the Trade Board Acts. For a discussion of their longevity, see Vivien Hart, *Bound by Our Constitution: Women, Workers, and the Minimum Wage* (Princeton, 1994), 61–62.

2. Ellis Hawley, *The New Deal and the Problem of Monopoly* (Princeton, 1966), 114.

3. Hart, *Bound by Our Constitution,* 20–21.

4. See complaints about group classifications in "Analysis of Protests against Laundry Code," Box 3077, National Recovery Administration Papers, Record Group 9, National Archives, Washington, D.C. (hereafter NRA Records).

5. Albert Johnson, ed., *Newsletter* 13 (23 June 1933): 4, Box 3084, NRA Records.

6. "Ministry of Labour Prosecution," *Cumberland Times* (30 April 1927) clipping in Lab 2/26603/TB 987, Ministry of Labour Records, Public Records Office, Kew Gardens (hereafter PRO).

7. See "Minute Sheet" (29 May 1925) Lab 2/TB 2938, PRO and examples that follow.

8. For American examples, see "Local Option Governs Laundry Code," 14 (March 1934): 66; "NRA Enters Enforcement Stage," *Laundry Age* 14 (1 May 1934): 50–52. For British examples, see "A Trade Board for the Laundry Industry," *The Power Laundry* (25 January 1919): 3; "The New Factories Bill," *The Power Laundry* (23 October 1926): 517.

9. See "Which Do You Choose?" and "What Is Your Answer?" *Laundry Age* 13 (1 June 1933): 14, 50.

10. "Minute Sheet," (23 April 1929) Lab 2/50240/TB 1646, PRO; "Note of Further Visits to the Empire Laundry," (April 1925) Lab 2/7068/TB 1068, PRO.

11. "The Laundry Code," *Laundry Age* 14 (1 March 1934): 72. This reprints the full text of the Laundry Code.

12. See, for instance, "Laundry Workers Organize!" (AFL pamphlet), Box 280, Women's Bureau Records, Record Group 86, National Archives, Washington, D.C. (hereafter WB Records).

13. J. E. Marran to Hugh Johnson, 12 September 1934, Box 3078, NRA Records.

14. "Sound Logic in Temporary Code," *Laundry Age* 13 (1 August 1933): 2.

15. W. Conover to J. C. Barringer, 28 June 1934, Box 3084, NRA Records.

16. C. T. Thomason to T. Lee Miller, 16 April 1934, Box 3077, NRA Records.

17. "Miss Collet's Evidence," Lab 2/1923/TB 102, PRO. Collet was alluding to the highly publicized campaigns of the Royal Society for the Protection of Animals.

18. "Minute Sheet," 27 November and 4 December 1937, TB 11/1949, PRO.

19. "Inspector's Report," (n.d.), TB 11/1949, PRO.

20. "Application for Exemption," (30 July 1933) Lab 2/1645, Case 5265, PRO.

21. "Report," Lab 2/26603/TB 987, PRO.

22. Hawley, *The New Deal and the Problem of Monopoly*, 247–69.

23. "State Code Puts New Jersey in the Spotlight," *Laundry Age* 14 (1 May 1934): 99–100; "Official Schedule," (12 December 1933) Box 3077, NRA Records; Memorandum by Henry Buckelew (23 January 1934) Box 3077, NRA Records; "Official Schedule of Minimum Retail Prices" (12 December 1933), Box 3077, NRA Records.

24. W. Conover to T. Barringer, 13 July 1934, Box 3084, NRA Records.

25. "Money Found for Minimum Wage Law Enforcement," *Laundry Age* 14 (1 July 1934): 90.

26. C. T. Thomason to Hugh Johnson, 18 September 1933, Box 3077, NRA Records.

27. "The Why of the Local Option," *The Power Laundry* 14 (1 May 1934): 91.

28. Handwritten notes on Chicago laundry study, n.d., Bertha Nienberg to Ethel Erickson, Box 280, WB Records.

29. James M. Hanley to Hugh Johnson, 7 August 1933, Box 3084, NRA Records.

30. W. W. Baker to Hugh Johnson, 30 June 1933, Box 3086, NRA Records.

31. W. Counsell, "Inspector's Report," (12 July 1921); Jameson to Dr. MacNamara, Minister of Labour, 27 September 1920, Lab 2/1310/TB 1645, PRO.

32. "Comments Arising out of the Investigation," (n.d.) Lab 2/1310/TB 1645, PRO.

33. "Memo," J. J. Stark to MacNamara, 14 October 1921, Lab 2/1310/TB 1645, PRO.

34. C. Jameson to Dr. MacNamara, 27 September 1921, Lab 2/1310/TB 1645, PRO.

35. *Laundry Journal* (20 August 1921): 6.

36. "Comments" (n.d.) Lab 2/1310/1645, PRO.

37. See "Clarke's Laundry," Lab 2/30053/TB 1645, PRO.

38. "Minute Sheet," Lab 2/45343/1645, PRO.

39. *Lydney Observer* (24 February 1928), clipping in Lab 2/45343/1645, PRO. *The Citizen* (24 February 1928), clipping in Lab 2/45343/1645, PRO.

40. J. J. Stark, "Milestones of Progress," *The Power Laundry* (4 May 1935): 653.

41. Frances Perkins, *The Roosevelt I Knew* (New York, 1946): 212.

10 / Women and Men and Unions

1. Rose Schneiderman with Lucy Goldthwaite, *All for One* (New York, 1967), 210, 217.

2. "The Laundry Workers Clean Up Their Industry," *Advance* (October 1937): 10; Charles Lionel Franklin, *The Negro Labor Unionist of New York: Problems and Conditions among Negroes in the Labor Unions in Manhattan with Special Reference to the N.R.A. and post–N.R.A. Situations* (New York, 1936), 180; *Advance* (February 1938): 20.

3. During the preceding twenty years, the AFL's ILWU had seldom reported more than six thousand members nationwide. U.S. Department of Labor, *Handbook of Labor Statistics*, 1929 edition, *Department of Labor Bulletin* 506 (Washington, D.C., 1929), 213; Estelle M. Stewart, *Handbook of American Trade-Unions*, 1936 edition, *Department of Labor Bulletin* 618 (Washington, D.C., 1936), 327.

4. Jean Collier Brown, "The Negro Woman Worker," *Women's Bureau Bulletin* 165 (1938), 15; Jacqueline Jones, *Labor of Love, Labor of Sorrow* (New York, 1985), 212.

5. Emma Crosby, "Conditions in Laundries," *New York Amsterdam News*, 4 April 1936, 12.

6. For a comparative overview, see Neville Kirk, *Labour and Society in Britain and the USA, Volume 2: Challenge and Accommodation, 1850–1939* (Aldershot, 1994).

7. See Norbert C. Soldon, *Women in British Trade Unions, 1874–1976* (Dublin, 1978), 103–6.

8. "Organizer's Report," (1925), reel 3, New York Women's Trade Union League Papers in *Papers of the Women's Trade Union League and Its Principle Leaders* (microfilm; Woodbridge, Conn., 1981), hereafter NYWTUL Papers.

9. "Organizer's Report–June 26, 1918–July 31, 1918," reel 2, NYWTUL Papers; "Report of the Work of the New York W.T.U.L. for February 1925," (26 February 1925) reel 3, NYWTUL Papers; Schneiderman, *All for One*, 213–14; "Report of Organization Work," (February 1928) reel 3, NYWTUL Papers; "Report of Organization Work," (March and April 1928) reel 3, NYWTUL Papers.

10. "Report of the Work of the New York W.T.U.L. for February 1925," (26 February 1925) reel 3, NYWTUL Papers.

11. See Irving Bernstein, *The Lean Years: A History of the American Worker, 1920–1933* (Baltimore, 1966).

12. Charles Lionel Franklin, *The Negro Labor Unionist of New York: Problems and Conditions among Negroes in the Labor Unions in Manhattan with Special Reference to the N.R.A. and post–N.R.A. Situations* (New York, 1936), 101, 106.

13. "Report of Organization Work" (March and April 1928) reel 3, NYWTUL Papers.

14. Stewart, *Handbook of American Trade-Unions,* 214.

15. Gary Edward Endelman, "Solidarity Forever: Rose Schneiderman and the Women's Trade Union League," (Ph.D. dissertation, University of Delaware, 1978), 156.

16. "Organization Report–January 8, 1934" reel 3, NYWTUL Papers.

17. Ibid.

18. "Interview with Mr. George Webber and Mr. Lundgren," (n.d.) Box 280, Women's Bureau Records, Record Group 86, National Archives, Washington D.C.

19. "Organization Report," (4 December 1933), reel 3, NYWTUL Papers.

20. "Organization Report," (8 January 1934), reel 3, NYWTUL Papers.

21. Ibid.

22. "Organization Report," (February 1934), reel 3, NYWTUL Papers.

23. Ibid.

24. Cover photograph, *Crisis* 41 (March 1934).

25. "Organization Report," (March 1934), reel 3, NYWTUL Papers.

26. Schneiderman, *All for One,* 216.

27. "Organization Report," (March 1934), reel 3, NYWTUL Papers.

28. Cited in Endelman, "Solidarity Forever," 59. Endelman's work is an excellent analysis of Schneiderman's motivations.

29. "Organization Report," (March 1936), reel 4, NYWTUL Papers. "Organization–1935–6," reel 4, NYWTUL Papers; "Organizer's Report," (1 April 1936–30 March 1937), reel 3, NYWTUL Papers.

30. Georgette Johnson, "Plight of Laundry Workers"; Marion Catherwood, "Minimum Wage Law," *New York Amsterdam News,* 21 March 1936, 12.

31. "Organizer's Report," (1 April 1936–30 March 1937), NYWTUL Papers, reel 4.

32. Sabina Martinez, "A Black Union Organizer," in *Black Women in White America*, ed. Gerda Lerner (New York, 1972), 263.

33. See, for example, "Striking Workers Tie Up Laundries," *New York Amsterdam News*, 3 April 1937, 1; "50 Strikers Try to Stop Scabs," *New York Amsterdam News*, 17 April 1937, 1; "150 Still Strike in Yonkers," *New York Amsterdam News*, 21 August 1937, 1.

34. "Executive Board Minutes," (23 January 1934), reel 3, NYWTUL Papers.

35. "Minutes of the Organizing Committee," (10 August 1937), reel 4, NYWTUL Papers.

36. After the breakup, an embittered Mackey waged a countercampaign against the CIO, distributing leaflets and filing suit. See "Minutes of Organization Committee," (10 August 1937), NYWTUL Papers. He had apparently caught wind of his impending loss, having written Sidney Hillman in April asking for an interview to discuss the CIO's intentions in regard to the laundry workers. In the letter, he praised the "wonderful" help he had received from Schneiderman and the league, and he also took credit for forming the joint committee. See Joseph Mackey to Sidney Hillman, 5 April 1937, Amalgamated Clothing Workers Papers, Box 77, Labor Management Research Center, Cornell University (hereafter ACW Papers). Hillman claimed to be too busy to see Mackey and sent him the address of a CIO representative in Washington D.C. See Sidney Hillman to Joseph Mackey, 7 April 1937, ACW Papers.

37. Schneiderman, *All for One*, 216–17.

38. "Accomplishments of the Laundry Workers Organizing Committee," *Advance* (May 1938): 15. In effect, Bernstein was already organizing laundry workers, because some manufacturers of "white duck" garments also offered relaundering services to hospitals, laboratories, and other businesses that bought their products. White duck was in fact a porous boundary between garment manufacturing and the laundry industry.

39. "Report of the Organizer," (September 1937), reel 4, NYWTUL Papers.

40. "The AFL Laundry Union's the Bunk," *Advance* (October 1939): 24; "Two Years of Laundry Labor Union, 1937–1939," *Advance* (August 1939): 20.

41. "Organization Work Speeds Ahead in Laundry Industry," *Advance* (January 1938): 12.

42. For an extensive analysis of the origins of Hillman's guiding philosophy, see Steven Fraser, *Labor Will Rule: Sidney Hillman and the Rise of American Labor* (New York, 1991).

43. The joint board consisted of representatives of various types of laundries—e.g., hand laundries, linen laundries, etc. In effect, this meant that the "leaders" of the industry ended up being on the board rather than a representative cross-sample. Jesse Thomas Carpenter, *Employers' Associations and Collective Bargaining in New York City* (Ithaca, 1950), 133–34.

44. See Michael Goldfield, "Race and the CIO: The Possibilities for Racial Egalitarianism during the 1930s and 1940s," *International and Working Class History* 44 (1993): 1–63; Ruth Milkman, "New Research in Women's Labor History," *Signs* 18 (1993): 376–88; Bruce Fehn, "Striking Women: Gender, Race, and Class in the United Packinghouse Workers of America (UPWA)," (Ph.D. dissertation, University of Wisconsin, Madison, 1991).

45. Although the basis of this strategy was in classic socialist ideology, it would be inaccurate to say that the ACWA was cultivating class consciousness. The ACWA's emphasis on both cooperation with employers and the stabilization of capitalism suggest a certain privileging of the needs of employers. Moreover, the bureaucratic structure of the ACWA was a cross-class alliance. The common thread was demonstrated allegiance to the labor movement and the ability to be part of trade union culture.

46. "A Laundry Worker's Creed," *Advance* (November 1938): 14.

47. "Ten Thousand Laundry Workers Get ACWA Contract," *Advance* (September 1937): 19.

48. "Minimum Wage Board," *Advance* (January 1938): 12.

49. "Laundry Workers Hours and Wages before Arbitrator," *Advance* (November 1937): 12.

50. "Union Completes Important Survey of Laundry Industry," *Advance* (February 1940): 23.

51. Goldfield, "Race and the CIO," 19.

52. "Minimum Wage Board," *Advance* (January 1938): 12. For an extensive analysis of gender imagery in CIO magazines, see Elizabeth Faue, *Community of Suffering and Struggle: Women, Men, and the Labor Movement in Minneapolis, 1915–1945* (Chapel Hill, N.C., 1991).

53. "Ironing Things Out," *Advance* (November 1937): 12.

54. "The Arbitration Court for the Laundry Industry," *Advance* (February 1938): 21; "Organization Work Speeds Ahead in the Laundry Industry," *Advance* (January 1938): 12.

55. "Joint Board, Local 300, United Laundry Workers," (photograph) *Advance* (November 1937): 12.

56. "Executive Board, Westchester County Local," *Advance* (April 1938): 12.

57. Sabina Martinez, "A Black Union Organizer," 264.

58. "To the General Executive Board of the Amalgamated Clothing Workers of America, Greetings!" (21 April 1939), 3, 18, Box 59, ACW Papers.

59. This was, of course, a strategy that proved problematic in the postwar era, as workers staged wildcat strikes in knowing defiance of labor-management accords. For an extended analysis, see George Lipsitz, *Rainbow at Midnight: Labor and Culture in the 1940s* (Urbana, Ill., 1994).

60. "Educational Department," *Advance* (December 1938): 14; "The Laundry Workers Joint Board Chorus," *Advance* (August 1938): 17; "Three LWJB Members in Broadway Hit Show" (January 1940): 21.

61. "Three Decades of Service," *New York State Department of Labor Bulletin* (1965): 19, Fiche 002,751, Schomburg Clipping File, New York.

62. Florence Rice, "It Takes a While to Realize That It Is Discrimination," *Black Women in White America*, 275–76.

63. "Evelyn Macon," *Rank and File*, ed. Ann Banks (New York, 1980), 127–28.

64. Ibid., 128.

65. "Re: Coverall Dry Cleaners Discharge of Kilbourn Clapsaddle," (4 March 1940), Box 2, Laundry and Dry Cleaning Drivers Local 360, Milwaukee Mss DT, Wisconsin Historical Society (hereafter LDCD).

66. Ibid.

67. Padway and Goldberg to Alois Mueller, 12 May 1942, Box 3, LDCD; Picketing Schedule (19 March 1940), Box 7, LDCD; Membership Lists 1936–1946, Box 8, LDCD.

68. Alois Mueller to Federated Trades Council, 2 February 1938, Box 1, LDCD.

69. For an example of a driver's contract, see Wisconsin Dye Works contract

with Clair Burfiend, 16 February 1937, Box 1, LDCD. See also "*Richard Planer v. Laundry and Dry Cleaning Drivers Local* before the Wisconsin Employment Relations Board," Box 2, LDCD.

70. George J. Ritchey to John M. Gillespie, 14 February 1938, Box 1, LDCD; Gillespie to Ritchey, 15 February 1938, Box 1, LDCD.

71. Alois Mueller to Members, 29 June 1938, Box 1, LDCD.

72. Harry Hillman to Alois Miller [sic], 3 October 1938, Local 360, Box 1, LDCD.

73. Oliver Adelman to Al Miller [sic], 26 September 1940, Box 2, LDCD.

74. Howard Fink to A. Miller [sic], 2 August 1939, Box 2, LDCD; Howard Fink to Julius P. Heil, 9 August 1939, Box 2, LDCD; Alois Mueller to Julius P. Heil, 11 August 1939, Box 2, LDCD.

75. Alois Mueller to John M. Gillespie, 22 November 1938, Box 1, LDCD; William Stadler to International Brotherhood of Teamsters, Executive Board, 25 November 1938, Box 1, LDCD; Alois Mueller to Members and Employers, 16 November 1938, Box 1, LDCD; Wood, Warner, and Tyrrell to Alois Mueller, 3 December 1938, Box 1, LDCD.

76. "Minutes of Executive Board," (12 September 1939), Box 8; Padway, Goldberg, and Tarrell to Alois Muller, 7 October 1939, Box 2, LDCD.

77. Alois Mueller to Members and Employers, 8 September 1938, Box 1, LDCD; Alois Mueller to City Beer Exchange, 11 March 1939, Box 2, LDCD.

78. "Minutes," (10 October 1939), Box 9, LDCD.

79. "Minutes," (9 February 1940), Box 9, LDCD.

80. The Entertainment Committee to Members' Wives, 10 March 1941, Box 2, LDCD.

81. John S. Picago to Fred Siewert, 22 June 1943, Box 3, LDCD.

82. George Scanlon to Elizabeth Mauer, 17 July 1945, Box 4, LDCD; Alois Mueller to Dorothy Fischer, 7 December 1942, Box 3, LDCD; George Scanlon to Franklin Schneider, 18 January 1946, Box 5, LDCD.

11 / Facing the Domestic Threat

1. Ruth Schwartz Cowan, *More Work for Mother: The Ironies of Household Technology from the Open Hearth to the Microwave* (New York, 1983), 107; Patricia Malcolmson, *English Laundresses: A Social History* (Urbana, Ill., 1986), 158.

2. U.S. Department of Commerce, Bureau of the Census, *Census of Manufactures: 1931, Power Laundries, Drycleaning and Redying Establishments* (Washington, D.C., 1933), 9. I have not been able to find similar figures for Great Britain.

3. See Sue Bowden and Avner Offer, "The Technological Revolution That Never Was: Gender, Class, and the Diffusion of Household Appliances in Interwar England," in *The Sex of Things: Gender and Consumption in Historical Perspective,* ed. Victoria de Grazia (Berkeley, 1996), 244–74; and Sue Bowden and Avner Offer, "Household Appliances and the Use of Time: The United States and Britain since the 1920s," *Economic History Review* 47 (November 1994): 745–46. Malcolmson, *English Laundresses,* 158–60, and Christine Zmroczek, "Women, Class, and Washing Machines, 1920s to 1960s," *Women's Studies International Forum* 15 (1992): 175, cite Cowan for American statistics.

4. Carolyn Haslett, *Household Electricity* (London, 1939), 78; Leslie Hannah, *Electrification before Nationalization: A Study of the Development of the Electricity Supply Industry in Britain to 1948* (Baltimore, 1979), 194–95, 208.

5. F. L. Hinchliff, "6% Net Profit Makes the Washer and Cleaner Business Always Important," *Electrical Merchandising* 43 (June 1930): 57. *Electrical Merchandising* published yearly sales figures compounded from data gathered from dealers. Although some exaggeration is undoubtedly present, the figures had to correlate roughly with the number of machines manufactured. For later years, see "Markets for Electrical Merchandise," *Electrical Merchandising* 51 (January 1934): 28, and "Statistical Summary," *Electrical Merchandising* 57 (January 1937): 9, which listed 1,533,300 electric and 199,000 gas machines sold.

6. Though no figures are available for receipts, comparing the increases in numbers of workers gives some indication of the difference. The number of American laundry workers doubled between 1920 and 1930, from 120,442 (men and women) to 240,520: U.S. Department of Commerce, Bureau of the Census, *Sixteenth Census of the United States: 1940,* vol. 3, *Population Comparative Occupational Statistics for the United States, 1870–1940* (Washington, D.C., 1943), 129. In Britain, there was a total of 108,637 laundry workers according to the 1921 British Census, 146,750 in 1931: "Eighteenth Annual Abstract of Labour, Statistics of the United Kingdom for 1921," Cmd. 2740 (1921), *Parliamentary Papers,* 15; "Twenty-First Annual Abstract of Labour," Cmd. 4625 (1931) *Parliamentary Papers,* 949. The 1931 figures also included workers in dyeing and drycleaning establishments not added into the 1921 numbers.

7. That survey counted 249,008 employees and gross receipts of $455,579,000.

U.S. Census, *Sixteenth Census of the United Sates: 1940,* vol. 3, *Service Establishments,* 443.

8. Ibid., vol. 3, *The Labor Force,* 79; Ibid., vol. 3, *Comparative Occupational Statistics,* 129.

9. Roi B. Woolley, "Ways That Sold and Are Selling . . . Washers," *Electrical Merchandising* 51 (May 1934): 48.

10. "Electrical Washing Day," *Electrical Age for Women* 1 (January 1927): 89.

11. Thomas P. Hughes, *Networks of Power: Electrification in Western Society, 1880–1930* (Baltimore, 1983), 128.

12. David Nye, *Electrifying America: Social Meanings of a New Technology, 1880–1940* (Cambridge, Mass., 1990), 261.

13. "Now 19,843,724 Wired Homes," *Electrical Merchandising* 51 (January 1934): 32.

14. Nye, *Electrifying America,* 319.

15. Pamela Sambrook, *Laundry Bygones* (Aylesbury, 1983), 11; T. A. B. Corley, *Domestic Electrical Appliances* (London, 1966), 27, 33. John Stevenson, *British Society, 1914–45* (Harmondsworth, 1984), 110.

16. Cowan, *More Work for Mother,* 94, 156; Nye, *Electrifying America,* 267.

17. Corley, *Domestic Electrical Appliances,* 34.

18. Nye, *Electrifying America,* 264. Like early washing machines, the first electric irons were quite expensive. In 1907, the Army and Navy Stores in London were advertising electric irons for 22 shillings. In comparison, a box iron could be had for less than 3 shillings. Sambrook, *Laundry Bygones,* 22.

19. William M. Emery, "Getting Back Home Laundry Business," *Electrical Merchandising* 43 (June 1930): 65. Nye, *Electrifying,* 18.

20. Sambrook, *Laundry Bygones,* 12.

21. "Just to Save a Few Pennies," *Laundry Age* (1 January 1936): 39.

22. For British examples see articles in the British version of *Good Housekeeping*: "A Lesson in Laundrywork," *Good Housekeeping* 10 (January 1927): 34–35; "Washing by Up-to-Date Methods," *Good Housekeeping* 9 (June 1926): 34–35. The primary purpose of the British journal *Electrical Age for Women* was to teach women how to use electrical appliances including washing machines and other home laundry equipment. See "Electrical Washing Day," *Electrical*

Age for Women 1 (January 1927): 88–89 and "A Modern Electric Laundry," 2 (July 1930) 16–17. For an American example, see Lydia Ray Balderston, *Laundering* (Philadelphia, 1923).

23. Roland Marchand, *Advertising the American Dream: Making Way for Modernity, 1920–1940* (Berkeley, 1985), 175. Heidi Irmgard Hartmann, "Capitalism and Women's Work in the Home, 1900–1930," (Ph.D. dissertation, Yale University, 1974), 319.

24. "Home Laundry Advantages," *Laundry Age* (1 May 1936): 44.

25. For a quantitative analysis of the changing patterns of consumption in the United States, see Martha L. Olney, *Buy Now, Pay Later: Advertising, Credit, and Consumer Durables in the 1920s* (Chapel Hill, N.C., 1991), 31.

26. "75 Cents Down," *Electrical Merchandising* 43 (May 1930): 59.

27. "The Menace of the Electric Home Washer," *The Power Laundry* (11 September 1926): 287.

28. Sambrook, *Laundry Bygones,* 11.

29. Laurence Wray, "The Merchandising Month: 'Competition with an Axe,'" *Electrical Merchandising* 53 (January 1935): 1.

30. Christine Zmroczek, "The Weekly Wash," in *This Working-Day World: Women's Lives and Culture(s) in Britain 1914–1945,* ed. Sybil Oldfield (London, 1994), 13.

31. Laurence Wary, "The Merchandising Month," *Electrical Merchandising* 57 (January 1937): 1.

32. G. E. Steadman, "For Salesmen Only," *Electrical Merchandising* 51 (January 1934): 46.

33. T. F. Blackburn, "Do's and Don't's of Coin Operated Washers," *Electrical Merchandising* 51 (January 1934): 44–45.

34. "Is There Actually Economy in Doing the Wash at Home? Della T. Lutes Says Yes!" *American Home* 11 (January 1934): 91.

35. Cowan, *More Work for Mother,* 150.

36. H. M. V. Advertisement, *Electrical Age for Women* 3 (October 1936): 158.

37. Jacob Swisher, "The Evolution of Washday," *Iowa Journal of History and Politics* 38 (January 1930): 30.

38. "The Zero Hour is Now!" *The Power Laundry* (13 July 1935): 41.

39. Susan Strasser, *Never Done: A History of American Housework* (New York, 1982), 120.

40. Nye, *Electrifying America*, 273.

41. "Glamour in the Laundry! Joy in Homemaking!" *Good Housekeeping* 104 (February 1937): 79.

42. "The 'Approved Laundry,'" *Good Housekeeping* 101 (October 1935): 93. Susan Strasser also suggests that the "girl in the white cap" was a widespread symbol of cleanliness, particularly well known from the publicity campaign of the Heinz corporation. Susan Strasser, *Satisfaction Guaranteed: The Making of the American Mass Market* (New York, 1989), 121.

43. "Are You Proud of Your Laundry?" *Good Housekeeping* 102 (May 1936): 87.

44. Quoted in Christina Hardyment, *From Mangle to Microwave: The Mechanization of Household Work* (Cambridge, 1988), 64.

45. "E. A. W. Luncheon," *Electrical Age for Women* (June 1926): 25. See also an article written by American home economist Christine Fredericks, "How the American Housewife Achieves Leisure through Electricity," *Electrical Age for Women* (April 1927): 98.

46. "Emotional Selling," *Electrical Merchandising* 52 (August 1934): [cover].

47. "Some Electrical 'Advice to the Lovelorn,'" *Electrical Merchandising* 33 (January 1925): 5035.

48. For a more extended analysis of the targeting of women by advertisers, see Glenna Matthews, *"Just a Housewife": The Rise and Fall of Domesticity in America* (New York, 1987), 188–89.

49. G. E. Steadman, "Something about Women," *Electrical Merchandising* 57 (June 1937): 4.

50. Ibid.

51. Roi B. Woolley, "Ways That Sold and Are Selling . . . Washers," *Electrical Merchandising* 51 (May 1934): 48.

52. "How to Get Your Share of 650,000 Washers This Year," *Electrical Merchandising* 33 (February 1925): 5151.

53. *Bundle News* (Madison, ca. 1930), Wisconsin Historical Society Pamphlet Collection.

54. Wilfrid L. Randell, *Electricity and Woman: Twenty-one Years of Progress* (London, 1945) 23.

55. "'Britain Now Leads' says Prominent Engineer, after Flying Trip to the States," *Laundry Record* (August 1935): 151.

56. Robert S. Lynd and Helen Merrell Lynd, *Middletown: A Study in American Culture* (New York, 1929), 175. The Lynds' analysis may be a product of sociologists' fascination with the power of advertising during the 1920s.

Bibliographical Essay

Rather than present an exhaustive survey, the following essay describes the most significant sources, both primary and secondary, that influenced the content and organization of this book. It is organized both topically and chronologically, to roughly coincide with the organization of the book itself.

General Works

Most secondary references to laundries and laundry workers are scattered through books and articles addressing other subjects. An exception is Patricia Malcolmson, *English Laundresses: A Social History, 1850–1930* (Urbana, Ill., 1986). Both British and American trade associations also commissioned industry histories. On Britain, see Ancliffe Prince, *The Craft of Laundering: A History of Fabric Cleaning* (London, 1970). On the United States, see Fred De Armond, *The Laundry Industry* (New York, 1950). Both should be used cautiously. For an analysis of the way the industry has presented its own history, see Arwen Mohun, "Laundrymen Construct Their World," *Technology and Culture* 36 (January 1997): 97–120.

The literature on the history of housework helps explain why women preferred not to do their own washing. Ruth Schwartz Cowan, *More Work for Mother: The Ironies of Household Technology from the Open Hearth to the Microwave* (New York, 1983), contrasts both domestic and industrial alternatives. On domestic laundrywork in the United States, see also Susan Strasser, *Never Done: A History of American Housework* (New York, 1982). Caroline Davidson, *A Woman's Work Is Never Done: A History of Housework in the British Isles, 1650–1950* (London, 1982), and Pamela Sambrook, *Laundry Bygones* (Aylesbury, 1983), are useful sources for Great Britain. Costume historians Christina Walkey and Vanda Foster offer additional important context in *Crinolines and Crimping Irons: Victorian Clothes: How They Were Cleaned and Cared For* (London, 1978).

Technological and Cultural Origins of Industry

This book posits the textile industry as one source of the first mechanized laundry equipment, suggesting Charles Strutt as one possible progenitor. R. S. Fitton and A. P. Wadsworth, *The Strutts and the Arkwrights, 1758–1830: A Study of the Early Factory System* (Manchester, 1958), remains the definitive work on the Strutts. For early textile technology, see Edward Baines, *A History of the Cotton Manufactures of Great Britain* (London, 1835). For background on patent records, see Christine Macleod, *Inventing the Industrial Revolution: The English Patent System, 1660–1800* (New York, 1988), and *Technology and Culture* (*Special Issue on Patents and Invention*) 32 (October 1991).

The laundry industry would not have been possible without an appropriate cultural context. Anthropologist Mary Douglas's *Purity and Danger: An Analysis of Concepts of Pollution and Taboo* (New York, 1966) provides the most important theoretical explanation for cultural definitions of cleanliness. See also Georges Vigarello, *Concepts of Cleanliness: Changing Attitudes in France since the Middle Ages*, trans. Jean Birrell (Cambridge, 1988); Alain Corbin, *The Foul and the Fragrant: Odor and the French Social Imagination* (Cambridge, 1986); and Norbert Elias's classic *The Civilizing Process*, vol. 1 of *The History of Manners* (New York, 1978).

Richard L. Bushman and Claudia L. Bushman, "The Early History of Cleanliness in America," *Journal of American History* 74 (1988), touches on eighteenth-century American ideas of cleanliness, as does Harold Donald Eberlein, "When Society First Took a Bath," in *Sickness and Health in America: Readings in the History of Medicine and Public Health*, ed. Judith Walzer Leavitt and Ronald L. Numbers (Madison, 1978). Most American accounts place the revolution in cleanliness in the period after the American Civil War. See Suellen Hoy's *Chasing Dirt: The American Pursuit of Cleanliness* (New York, 1995).

I put the transformation earlier, in the late eighteenth and early nineteenth century: an interpretation that coincides more closely with the literature on the history of public health and the challenges of urbanization. For the United States, begin with John Duffy, *The Sanitarians: A History of American Public Health* (Urbana, Ill., 1990). On Britain, see Anthony S. Wohl, *Endangered Lives: Public Health in Victorian Britain* (Cambridge, Mass., 1983); and Roy Porter, "Cleaning up the Great Wen: Public Health in Eighteenth-Century London," in

Living and Dying in London, ed. W. F. Bynum and Roy Porter (London, 1991). Nelson Manfred Blake, *Water for the Cities: A History of the Urban Water Supply Problem in the United States* (Syracuse, N.Y., 1956), is the standard reference for the growth of urban water systems in the United States.

Prescriptive and descriptive passages on cleanliness are widespread in eighteenth- and nineteenth-century etiquette manuals, novels, sermons, and other primary sources. The British Library has cataloged many of its sermons by subject. Travel literature from this period is especially rich in observations about hygenic practices. For an extensive bibliography of etiquette manuals, see John F. Kasson, *Rudeness and Civility: Manners in Nineteenth-Century Urban America* (New York, 1990).

Laundries as Businesses

Despite occasional calls for more attention to the history of small businesses, for instance, Mansel G. Blackford, "Surveys and Debates in Small Business in America: A Historiographical Survey," *Business History Review* 65 (spring 1991): 1–26, most of the business history literature remains focused on big businesses. The work of Philip Scranton on batch production has, however, had some effect in turning the field away from its emphasis on large-scale, mass-production industries. See Philip Scranton, "Diversity in Diversity: Flexible Production and American Industrialization, 1880–1930," *Business History Review* 65 (spring 1991): 27–90. The most relevant secondary literature I found was on small retailers and their trade associations. See, in particular, David Monod, "Culture without Class: Canada's Retailers and the Problem of Group Identification, 1890–1940," *Journal of Social History* 23 (spring 1995): 521–45; and Chris Hosgood, "A 'Brave and Daring Folk'? Shopkeepers and Trade Associations in Victorian and Edwardian England," *Journal of Social History* 26 (winter 1992): 285–308.

The nature of laundry businesses was not conducive to generating or saving records. I have been unable to locate anything more substantial than account books for American laundries established before the 1920s. However, some British regional archives have preserved the records of a few laundries that functioned as long-lived joint-stock companies. This account relies on the extensive records of Bird's Laundry, housed in the Tyne and Wear Archives, and those of the Millbay Laundry, now at the Devon County Record Office. A few British laundry owners have also commissioned histories of their companies. See Eric B. Dobson, *Hartonclean, 1884–1984* (Newcastle-upon-Tyne, 1983); David G. Thomas and Brenda J. Sowan, *The Walton Lodge Sanitary Laundry* (London, 1977); and Cecil Roberts, *The First Fifty Years of "Millbay"* (Plymouth, 1948).

The laundry industry did, however, support a large trade literature. Before the 1920s, the *National Laundry Journal* (first published in 1884) represented the Laundrymen's National Association and was the main journal in the United States. Its longest-lived competitor was the *American Laundry Journal.* See also a host of shorter-lived and regional publications, most notably the *Pacific Laundryman.* The first and most important British periodical, the *Laundry Journal,* began in 1885, followed by the *Laundry Record,* which gradually replaced it in importance. In America, after 1920, the *National Laundry Journal* was gradually eclipsed in importance by *Laundry Age,* while in Britain, the *Laundry Record* was similarly displaced by the *Power Laundry* after the *Laundry Record's* editor decided to use its pages to promote evangelical Christianity.

While full runs of the British journals are available in the British Library, no copies of the nineteenth-century volumes of American journals seem to have survived in public collections. However, selected articles from the *National Laundry Journal* and other publications were reprinted as books. See, in particular, *The Modern Laundry Guide: A Collection of the Best Articles Published in the National Laundry Journal* (Chicago, 1905), and C. A. Royce, *The Steam Laundry and Its Methods: Essays Read at the Laundrymen's National Conventions* (Chicago, 1894).

Publishers of laundry journals and various entrepreneurs also published guides for laundry owners and managers. Chicago publishers created a small library of books such as the National Laundry Journal, *The Modern Laundry Guide* (Chicago, 1905); C. A. Royce, *The Steam Laundry and Its Methods.* British titles include *Laundry Management: A Handbook for Use in Private and Public Laundries . . . by the Editor of the Laundry Journal* (London, 1889); *The Laundry Trade Directory and Launderer's Handbook with Diary* (London, 1908), C. Cadogan Rothery and H. O. Edmonds, *The Modern Steam Laundry: Its Construction, Equipment and Management* (London, 1909), 2 vols.

Readers interested in the workings of specific pieces of machinery can also consult the substantial number of surviving British and American trade catalogs. During the period covered by this study, the Troy Laundry Machinery Company dominated the industry. Its many catalogues are scattered through both British and American libraries, overshadowing catalogs from many smaller companies, such as the Adams Laundry Machinery Company. While Troy also exported to Britain, W. Summerscales and Sons was the largest British manufacturer. See, for instance, *W. Summerscales and Sons, Ltd.: Makers of High-Class Laundry Machinery* (Keighly, 1897).

Work, Workers, and Work Process

For a discussion of relevant secondary literature on women and work, see notes 3 and 4 to the introduction of this volume. On the functions and dynamics of skill in laundries, see also Arwen Mohun, "Why Mrs. Harrison Never Learned to Iron: Gender, Skill, and Mechanization in the American Steam Laundry Industry," *Gender and History* 8 (August 1996): 231–51.

Laundry workers left even fewer written records behind than their employers. This book relies primarily on government reports and the writings of reformers and reform groups. A few British working-class biographies do deal with laundry work, including Flora Thompson, *Lark Rise to Candleford* (London, 1984); Alice Linton, *Not Expecting Miracles* (London, 1982); Kathleen Woodward, *Jipping Street: Childhood in a London Slum* (London, 1928). American narratives, such as Cynthia Woods, "You Have to Fight for Freedom," in *Rank and File: Personal Histories by Working-Class Organizers*, ed. Alice Lynd and Staughton Lynd (New York, 1988), describe the later interwar period.

Very little material of any sort survives on workers and workplaces before the 1890s. Most of what can be found deals with washerwomen who either took in laundry or worked in hand laundries. Representative sources include William Burns, *Female Life in New York City Illustrated with Forty-four Portraits from Real Life* (Philadelphia, 1859) and Henry Mayhew, *London Labour and the London Poor*, vol. 2 (London, 1861).

As the industry grew and laundries became emblematic of the problem of women's employment in industry, various government agencies commissioned reports and held hearings. For Britain, begin with reports by royal commissions and factory inspectors in the Parliamentary Papers (organized by number as "command papers"), particularly the "Royal Commission on Labour: Employment of Women in Laundries," C. 6894 (1883–94); "Annual Report on the Work of H. M. Women Inspectors During 1897," C. 8965 (1898); "Report on the Work of H. M. Women Inspectors of Factories," Cd. 223 (1900); and "Report of Principal Lady Inspector," C. 223 (1899). Unfortunately no manuscript material survives from the factory inspectors for this period. Researchers interested in the effect of war mobilization on the industry should consult "Women in Industry: Report of the War Cabinet Committee on Women in Industry," Cmd. 135 (1919).

After the turn of the century, both the United States federal government and individual states published many reports on laundry workers. A useful starting place is U.S. Senate, *Report on Woman and Child Wage Earners in the United States*, vol. 12, *Employment of Women in Laundries*, 61st Cong., 2d Sess., S. Doc.

645. For state-level reports, see, for instance, New York Department of Labor, *Working Conditions in New York Steam Laundries* (Albany, 1912); and New York Department of Labor, "A Study of Hygienic Conditions in Steam Laundries and Their Effect Upon the Health of Workers," *Special Bulletin No. 130* (Albany, 1924). U.S. Department of Labor, Bureau of Labor Statistics, *Employment of Women in Power Laundries in Milwaukee* (Washington, D.C., 1913) is also particularly useful.

The 1890s also gave rise to a transatlantic tradition in investigative sociology. Many investigators took laundry workers as one of their subjects. For studies by these investigators in Britain, see Charles Booth, *Life and Labour of the People of London*, 2d. ser., vol. 8 (New York, 1896); George R. Sims, *Off the Track in London* (London, 1911); Thomas Oliver, ed., *The Dangerous Trades: The Historical, Social, and Legal Aspects of Industrial Occupations as Affecting Health by a Number of Experts* (London, 1902); and the Women's Industrial Council, *Annual Report*, 1905–1906 (London, 1906).

Elizabeth Beardsley Butler, *Women and the Trades: Pittsburgh, 1907–1908* (1909; reprint, Pittsburgh, 1984), contains a wealth of information on workers and work processes in Pittsburgh laundries. See also Katherine Anthony, *Mothers Who Must Earn* (New York, 1914), and Sue Ainslie Clark and Edith Wyatt, *Making Both Ends Meet: The Income and Outlay of New York Working Girls* (New York, 1911). Some muckraking journalists also disguised themselves as laundry workers and wrote about their experiences, notably Rheta Childe Dorr, *A Woman of Fifty* (New York, 1924).

The progressive reformers who staffed the newly established Women's Bureau of the United States Department of Labor after World War I brought with them a particular interest in laundries. See Ethel L. Best and Ethel Erickson, *A Survey of Laundries and Their Women Workers in Twenty-three Cities* (Washington, D.C., 1930); Bertha M. Nienberg and Bertha Blair, "Factors Affecting Wages in Power Laundries," *Bulletin of the Women's Bureau* 14 (1936); and Mary Elizabeth Pidgeon, "Women in the Economy of the United States of America: A Summary Report," *Women's Bureau Bulletin* 155 (1937). Very detailed information can also be found in the individual schedules collected by investigators, now part of the Women's Bureau Papers, Record Group 86, National Archives, Washington, D.C.

The sources cited above contain information on African American workers as well. Also worth consulting are Jean Collier Brown, "The Negro Woman Worker," *Women's Bureau Bulletin* 165 (1938); Mary Van Kleeck, "New Standards for Negro Women in Industry," *Life and Labor* (June 1919); Jane Filley, *Consider the Laundry Workers* (New York, 1937); Chicago Commission on Race Relations, *The Negro in Chicago: A Study of Race Relations and a Race Riot* (Chicago, 1923).

Unionization and Organization

With the exception of the collar laundresses described in Carole Turbin, *Working Women of Collar City: Gender, Class, and Community in Troy, 1864–86* (Urbana, Ill., 1992), and the African American washerwomen, covered by Tera Hunter in *To 'Joy My Freedom: Southern Black Women's Lives and Labor After the Civil War* (Cambridge, Mass., 1997), very little information exists on organization amongst laundry workers before the 1890 in either the United States or Great Britain. The Knights of Labor did include a number of laundryworkers' locals, although most seem to have been very short-lived. See Jonathan Garlock, *Guide to the Local Assemblies of the Knights of Labor* (Westport, Conn., 1983). For more on early organizations, consult John B. Andrews and W. D. P. Bliss, *History of Women in Trade Unions* (New York, 1974); and Ruth Delzell, *Early History of Women Trade Unionists of America* (Chicago, 1919). For Britain, see Norbert Soldon, *Women in British Trade Unions, 1874–1976* (Totowa, N.J., 1978); and Norbert Soldon, "British Women and Trade Unionism: Opportunities Made and Missed," in *The World of Women's Trade Unionism: Comparative Historical Essays*, ed. Norbert Soldon (London, 1985); Barbara Drake, *Women in Trade Unions* (London, 1984). A comparative overview of labor organization can be found in Neville Kirk, *Labour and Society in Britain and the USA, Volume 2: Challenge and Accommodation, 1850–1939* (Aldershot, 1994).

For both Britain and the United States, the records of the Women's Trade Union League (in its various national and regional incarnations) provide the most substantial documentation of efforts to organize laundry workers and to pass legislation pertaining to their trade. Robin Miller Jacoby's *The British and American Women's Trade Union Leagues* (Brooklyn, 1994) provided a particularly useful comparative perspective. This account also utilizes Nancy Schrom Dye, *As Equals and as Sisters: Feminism, the Labor Movement, and the Women's Trade Union League of New York* (Columbia, Mo., 1980), but challenges her periodization by following the Leagues' activities into the 1930s. On the participants in the British League, see also Philipa Levine, *Victorian Feminism, 1850–1900* (Tallahassee, Fla., 1987).

The League's involvement with British laundry workers is partially documented in the organization's annual reports included in the *Women's Trade Union League Papers* (microfilm; Wakefield, West Yorkshire, 1975) as well as in its affiliated journals: the *Women's Union Journal*, the *Women's Trade Union Review*, and the *Englishwoman's Review*.

The American League was created in 1906, long after its British counterpart, and was organized differently, with regional offices as well as a national headquarters. The surviving papers of all the major chapters, as well as the papers of principal leaders, are available together on microfilm: *Papers of the Women's Trade Union League and Its Principle Leaders* (microfilm; Woodbridge, Conn., 1981). The New York League papers are the most complete and useful for understanding the laundry story. For the interwar period, see also the League's journal *Life and Labor*.

In contrast to unions in other major American cities, the San Francisco local laundry unions did not have a strong connection to the WTUL. For this book, I have heavily relied upon Lillian Ruth Mathews, *Women in Trade Unions in San Francisco* (Berkeley, 1913). On the Anti-Jap Laundry League see also Robert M. Robinson, "A History of the Teamsters in the San Francisco Bay Area, 1850–1950" (Ph.D. dissertation, University of California, Berkeley, 1951); and Anti-Jap Laundry League, *Pacific Coast Convention of the Anti-Jap Laundry League* (San Francisco, 1908). Extensive records for the Laundry and Dry Cleaning Drivers Local 360 are held at the Wisconsin Historical Society.

On organizing African American laundry workers during the interwar period, consult the papers of the New York WTUL; Charles Lionel Franklin, *The Negro Labor Unionist of New York: Problems and Conditions among Negroes in the Labor Unions in Manhattan with Special Reference to the N.R.A. and post–N.R.A. Situations* (New York, 1936); and the *New York Amsterdam News*–the most important newspaper for the New York African American community. On the ACWA union, see the *Advance* and the Amalgamated Clothing Workers Papers, Box 77, Labor Management Research Center, Cornell University.

Both CIO unionism in general and the ACWA in particular have received substantial attention from historians. Steven Fraser, *Labor Will Rule: Sidney Hillman and the Rise of American Labor* (New York, 1991), analyzes the origins of Hillman's guiding philosophy in shaping the ACWA. Jesse Thomas Carpenter, *Employers' Associations and Collective Bargaining in New York City* (Ithaca, 1950), remains a useful guide to the context in which the ACWA laundry worker's union functioned. The contentious debate amongst scholars about the race and gender politics of the CIO is also particularly relevant to this subject. See Michael Goldfield, "Race and the CIO: The Possibilities for Racial Egalitarianism during the 1930s and 1940s," *International and Working Class History* 44 (1993): 1–63; Ruth Milkman, "New Research in Women's Labor History," *Signs*

18 (1993): 376–88; Bruce Fehn, "Striking Women: Gender, Race, and Class in the United Packinghouse Workers of America (UPWA)," (Ph.D. dissertation, University of Wisconsin, Madison, 1991); and Elizabeth Faue, *Community of Suffering and Struggle: Women, Men, and the Labor Movement in Minneapolis, 1915–1945* (Chapel Hill, N.C., 1991).

See also biographies and autobiographies of women reformers and labor leaders including Josephine Goldmark, *Impatient Crusader: Florence Kelley's Life Story* (Urbana, Ill., 1953); Rose Squire, *Thirty Years in the Public Service: An Industrial Retrospective* (London, 1927); Agnes Nestor, *Women's Trade Union Leader* (Rockford, Ill., 1954); Rose Schneiderman with Lucy Goldthwaite, *All for One* (New York, 1967); and Gary Edward Endelman, "Solidarity Forever: Rose Schneiderman and the Women's Trade Union League," (Ph.D. dissertation, University of Delaware, 1978).

The historical struggle over protective legislation for women has inspired an enormous secondary literature. Unlike many of the other topics dealt with in this book, some of the recent literature in this area is comparative. Begin with Vivien Hart, *Bound by Our Constitution: Women, Workers, and the Minimum Wage* (Princeton, 1994). See also Ulla Wikander, Alice Kessler-Harris, and Jane Lewis, eds., *Protecting Women: Labor Legislation in Europe, the United States, and Australia, 1880–1920* (Urbana, Ill., 1995); Theda Skocpol, *Protecting Soldiers and Mothers: The Political Origins of Social Policy in the United States* (Cambridge, Mass., 1992); and Linda Gordon, "The New Feminist Scholarship on the Welfare State," in *Women, the State, and Welfare*, ed. Linda Gordon (Madison, 1990).

On the WTUL and protective legislation in Britain, see Rosemary Feurer, "The Meaning of 'Sisterhood': The British Women's Movement and Protective Labor Legislation, 1870–1900," *Victorian Studies* 31 (winter 1988): 234; and Jacoby, *British and American Women's Trade Union Leagues.* Malcolmson, *English Laundresses,* provides a detailed analysis of the debate between pro- and anti-protectionist feminists, specifically over the inclusion of laundries under the Factory Acts.

On inclusion of laundries under the acts, see Parliamentary debates as well as Gertrude Tuckwell, "The More Obvious Defects in Our Factory Code," in *The Case for the Factory Acts,* ed. Beatrice Webb (London, 1901); Helen Bosanquet, Louise Creighton, and Beatrice Webb, "Law and the Laundry," *Nineteenth Century* 41 (1897): 225; Gertrude Tuckwell, "The Government Factory Bill of 1900," *Fortnightly Review* new ser. 73 (1900): 972; and Margaret McDonald, *Labour Laws for Women: Their Reason and Their Result* (London, 1900).

Mary Drake McFeely, *Lady Inspectors: The Campaign for a Better Workplace, 1893–1921* (Oxford, 1988) describes the involvement of female factory inspectors in the British laundry industry. Also useful are Squire, *Thirty Years in the Public Service*; B. Leigh Hutchins and Amy Harrison, *A History of Factory Inspection* (New York, 1970); and Adelaide Anderson, *Women in the Factory: An Administrative Adventure, 1893–1921* (New York, 1922).

There is no British equivalent to the central historiographical importance of *Muller v. Oregon*. For an overview of the scholarly debates on this case, begin with Nancy Woloch, *Muller v. Oregon: A Brief History with Documents* (Boston, 1996). See also Candice Dalrymple, *Sexual Distinctions in the Law: Early Maximum Hour Decisions of the United States Supreme Court, 1905–1917* (New York, 1987); Nancy S. Erickson, "*Muller v. Oregon* Reconsidered: The Origin of a Sex-Based Doctrine of Liberty of Contract," in *Labor History* 30 (spring 1989): 228–50; Susan Lehrer, *Origins of Protective Legislation for Women, 1905–1925* (Albany, 1987); Alice Kessler-Harris, "Law and a Living: The Gendered Content of 'Free Labor,'" in *Gender, Class, Race, and Reform in the Progressive Era*, ed. Noralee Frankel and Nancy S. Dye (Lexington, Ky., 1991), 87–109.

In contrast to this huge body of commentary, almost no primary source material regarding *Muller* has survived, other than legal documents and the accounts of the participants. Most scholars continue to rely on the version in Goldmark, *Impatient Crusader*. Researchers interested in Kurt Muller and the laundry aspects of the case should consult accounts in the Portland *Oregonian*.

Hart, *Bound by Our Constitution*, is the most comprehensive secondary treatment of the Trade Board Acts. See also Sheila Blackburn, "Ideology and Social Policy: The Origins of the Trade Board Act," *Historical Journal* 34 (March 1991): 43–64. Dorothy Sells's two books on the acts, the second written for a New Deal audience, are also invaluable sources: Dorothy Sells, *The British Trade Board System* (London, 1923); and Sells, *British Wage Boards: A Study in Industrial Democracy* (Washington, D.C., 1939). Ministry of Labour records on the creation and administration of the Acts are a treasure trove of virtually untouched materials.

Debates over the historical meaning of the NRA has generated a substantial historical literature, most of which is focused on Gabriel Kolko's "triumph of conservativism" argument. See, in particular, Colin Gordon, *New Deals: Business, Labor, and Politics in America, 1920–1935* (London, 1994); Ellis Hawley, *The New Deal and the Problem of Monopoly* (Princeton, 1966); and Robert F. Himmelberg, *The Origins of the National Recovery Administration: Business, Government, and the Trade Association Issue, 1921–1933*, 2d ed. (New York, 1993). My findings suggest that this interpretation overlooks the international context in which the NRA was constructed. For a similar critique, see Morton Keller, *Regulating a New Economy: Public Policy and Economic Change in America, 1900–1933* (Cambridge, 1990). For more on the racial politics of the NRA, begin with John Kirby, *The New Deal Era and Blacks: A Study of Black and White Race Thought, 1933–1945* (Urbana, Ill., 1971). My book's account of the NRA and the laundry industry relies primarily on NRA records held by the National Archives in Washington, D.C.

Washing Machines and Laundries

Despite more than twenty years of scholarship on domestic technology, the secondary literature on the adoption of washing machines in Britain and the United States remains empirically weak. Pioneering works like Cowan, *More Work for Mother,* Strasser, *Never Done,* and, to a lesser extent, Heidi Irmgard Hartmann, "Capitalism and Women's Work in the Home, 1900–1930," (Ph.D. dissertation, Yale University, 1974) have not been adequately reexamined or researched in more recent scholarship. Rigorous quantitative analysis, in particular, remains to be done.

Recent literature on British patterns includes Christina Hardyment, *From Mangle to Microwave: The Mechanization of Household Work* (Cambridge, 1988); Sue Bowden and Avner Offer, "The Technological Revolution that Never Was: Gender, Class, and the Diffusion of Household Appliances in Interwar England," in *The Sex of Things: Gender and Consumption in Historical Perspective,* ed. Victoria de Grazia (Berkeley, 1996), 244–74; Bowden and Offer, "Household Appliances and the Use of Time"; and Christine Zmroczek, "Women, Class, and Washing Machines, 1920s to 1960s," *Women's Studies International Forum* 15 (1992): 175.

For more general information on electrification and consumption in the United States, see David Nye, *Electrifying America: Social Meanings of a New Technology, 1880–1940* (Cambridge, Mass., 1990), Ronald Tobey, *Technology as Freedom: The New Deal and the Electrical Modernization of the American Home* (Berkeley, 1996). On Britain, see T. A. B. Corley, *Domestic Electrical Appliances* (London, 1966); and Leslie Hannah, *Electrification before Nationalization: A Study of the Development of the Electricity Supply Industry in Britain to 1948* (Baltimore, 1979).

For information about patterns of marketing and consumption, I have relied upon census data and trade journals, including laundry industry journals and *Electrical Merchandising,* the principal American trade journal for electrical appliance retailers. The publications and archives of the Electrical Association for Women in Great Britain also provide insight into the adoption of electrical technologies in Britain. See issues of the journal *Electrical Age for Women*; Carolyn Haslett, *Household Electricity* (London, 1939); and Wilfrid L. Randell, *Electricity and Woman: Twenty-one Years of Progress* (London, 1945).

Many types of laundries lie outside the main focus of this book. For more on Chinese laundries, see Paul C. Siu, *The Chinese Laundryman: A Study in Social Isolation* (New York, 1987) and Renquiu Yu, *To Save China, To Save Ourselves: The Chinese Hand Laundry Alliance of New York* (Philadelphia, 1992). I know of no substantive work on charity laundries, however the Charity Service Society Archives, at the Columbia University Rare Books and Manuscripts Library, New York, contains excellent records on a New York charity laundry run by the Charity Organization Society. See also Susan J. Kleinberg, *The Shadow of the Mills: Working-Class Families in Pittsburgh* (Pittsburgh, 1989), on Pittsburgh charity laundries. The laundries run by hotels, hospitals, and manufacturers also have histories worthy of further attention.

Index

Library of Congress Cataloging-in-Publication Data

Mohun, Arwen, 1961–

Steam laundries : gender, technology, and work in the United States and Great Britain, 1880–1940 / Arwen P. Mohun

p. cm. – (Johns Hopkins studies in the history of technology; new ser., no. 25)

Includes bibliographical references and index.

ISBN 0-8018-6002-4 (alk. paper)

1. Laundry industry–United States–History. 2. Laundry industry–Great Britain–History. I. Title. II. Series.

HD9999.L383U634 1999 98-38245

338.7'6166713'0973–dc21 CIP

Printed in the United States
781500004B